Introduction To

The Quantum Theory Of Matter and Radiation

A Brief History Of The Exponential Growth Of
Radical Ideas That Revolutionized Physics

By

Mohamed F. El-Hewie

Latest Revision:

January 31, 2012

TABLE OF CONTENTS

Page

PREFACE   4
The Three Stages of Development of
Quantum Mechanics
The Challenges to Quantum Mechanics   4

CHAPTER 1:
THE ROAD TO THE ATOM: 1792-1900 6

The Creation From Nothing   6
The Road To Conservation   6
Newton Laws Of Conservation   8
Beginning Of The Steam Age   9
The French Revolution   10
Birth Of Electricity   13
Nobel's Incentive   17
Landmarks of the Nineteenth Century   17
Ethereal Space   19
Light Spectrum of Elements   19
Discovery Of The Electron   20
Thomson Model Of The Atom   25
Discovery Of X-rays   26
Discovery Of Radioactivity   26
Classical Senses and Canons   27
Birth of Quantum Mechanics   28
Thermal Radiation   28
Black Body Radiation   29
Boltzmann-Wien Laws   30
Maxwellian distribution of molecular velocities   30
Boltzmann's theorem on entropy and probability   31
Equilibrium conditions and distribution law   35
Rayleigh-Jeans Law   37

CHAPTER 2:
THE BIRTH OF THE QUANTA   39
Thermal Radiation at High Frequencies   39
The Quanta   40
Photons And Corpuscles   41
Lorentz Transformation   43
Discovery Of Photons   44
Dual Nature of Light   45
Rutherford Atom Model   46
Bohr Atom Model   48
Privileged Atomic Quantum Orbits   49
Emission And Absorption Of Radiation   52
Brightness And Spin Of Emitted Radiation   53
de Broglie Matter Waves   54
Dual Properties of Electrons   57
Electron Diffraction   58
Quantum Wave Theory   60
Probability Wave Function   61
Uncertainty Relation   62
Tunnel Effect   63
Schrödinger Equation   64
Harmonic Oscillator   67
Pauli's Exclusion Principle   69
The Spin   70
Electronic Architecture of the Atom   71
Periodic Table of Elements   75
Atomic Spectrum   76

Page

CHAPTER 3:
MOLECULES AND SOLIDS   79
Molecules   79
Ionic molecule   79
Covalent molecules   79
Quantum Mechanics of Solids   80
Compound Crystals   80
Metallic Crystals   80
Stark Effect   81
Molecular Levels of Energy   82
Insulators   82
Conduction In Metal   83
Superconductivity   83
Semiconductors   84

CHAPTER 4:
THE INTERIOR OF THE NUCLEUS   87
Radioactivity   87
The Nucleus Of The Atom   87
The Neutron   89
Nuclear Forces   89
Discovery Of The Meson   91
Nuclear Stability And Abundance   91
Alpha Particle Decay   94
Nuclear Energy Levels   94
Gamma Ray Decay   96
Nuclear Fission   96
Beta Particles Decay   98
The Discovery Of The Neutrino   100
Electron Capture   101

CHAPTER 5:
ELEMENTARY NUCLEAR PARTICLES   102
Relativistic Quantum Mechanics   103
Dirac's Quantum Wave Equations   105
Discovery Of Anti-Matter   106

Action At A Distance   109
Field And Matter   112
Decay of Subatomic Particles   113
Deduction Of Coulomb's Law   116
Classification Of Elementary Particles   118
Decay Of Elementary Particles   122

CHAPTER 6: CRITIQUE ON
QUANTUM MECHANICS   127

Critique On The Special Theory Of Relativity   127
Critique On The Theory Of Matter Waves   129

CHAPTER 7: CONTRIBUTION
BY NATIONS OF SCIENTISTS   131

CHAPTER 8:
HIGHLIGHTS OF MAIN FACTS   133

REFERENCES   136

PREFACE

Man's curiosity in the essence of matter has never waned since the beginning of recorded history. The Ancient Greek believed that the entire universe consisted of four elements *water, air, earth, and fire*. The four elements act over hundreds and thousands of years to pulverize matter into dust. The limit of the size of such reduction was well known by the ancient philosophers Epicurus, Democritus and others. These particles were given the name "atoms", meaning indivisible in Greek. Their chief property was that no further division is possible.

This book narrates the brief history of the people, the events, and the ideas that refined our knowledge of the ancient elements of nature and that advanced mankind from reliance of the horse-drawn wagon into the electrical, nuclear, and space ages. The book describes the transformation of the classical perception of matter and wave into the new world of quantum mechanics. That comprises the efforts made to decipher the governing laws of the structure of atoms, molecules, crystals and elementary particles. The newly discovered conservation laws led to the discovery of some of the fundamental properties of the interaction of particles and fields. Further, the book analyses the modifications that assisted Quantum Mechanics in the discovery of new subatomic particles which in turn led to the weakening of the Quantum Theory and pressed the need for a new approach of the understanding of the behaviors of the elementary nuclear particles.

During the hundred and ten odd years of its existence, Quantum Mechanics has passed through *three distinct stages of development:*

Epicurus (341 BCE -270 BCE) believed that the events which occurred on earth were ultimately based on the motions and interactions of atoms moving in empty space. He is credited for discovering the first scientific method that held that nothing should be believed unless it has been tested and witnessed.

First Stage: Electronic Structure Of The Atom
This began by the discovery of Planck's quanta and ended by the discovery of de Broglie's matter waves. This period could be properly described as the stage of *structuring the electronic shells of the atom*. It embraced 25 years-from the discovery of the material properties of light waves to the discovery of the wave properties of material particles. During those years, Einstein and Bohr advanced new theories for electrons and photons that remedied the deficit of Classical Mechanics in dealing with subatomic phenomena. Both scientists acknowledged the severe limitation of classical conservation in dealing with the interrelation of wave and matter. Einstein questioned the classical constancy of mass and time and proposed an inhomogeneous space distorted by the presence of mass. Bohr went farther to set new rules for the motion electrons in steady states of energy levels unknown to Classical Mechanics.

Second Stage: Nuclear Forces Of The Atom
This began with de Broglie's discovery of matter waves, in 1924 and ended with the end of WWII. It could be properly described as the stage of deciphering *the nuclear forces of the atomic core*. Despite its extreme mathematical simplicity, de Broglie's matter waves glued many scattered pieces of the subatomic puzzle in such manner that Bohr's atomic energy-states could be rendered to mathematical modeling by Maxwell's wave equations. During this stage, and within the exceptionally short period of about 5 years, Schrödinger and Heisenberg created the basic framework of the Quantum Mechanics. The two scientists built their mathematical model on the then available old and new laws and

experimental data. They started by Coulomb's potential, classical conservation of energy and momentum, Planck's new quanta, and Bohr-Rutherford's configuration of the atomic particles and orbits. However, the new quantum wave equation skipped the special relativity relations. In 1928, Dirac re-synthesized Quantum Mechanics and Einstein's special relativity into a more powerful mathematical wave equation. Rutherford's experimental transformation of elements between 1911 and 1932 gave birth to nuclear fission. The theory of the atomic nucleus was created and new elementary particles were discovered that revealed some of the secrets of the nuclear forces.

Atomic nuclei were conceived around 1911. Twenty years later, in 1932 to be exact, the particles that make up the nucleus were defined and the forces operative between nuclear particles were discovered and explained. Another thirteen years later, the atomic age was thrust at full speed in the form of atomic bombs and nuclear reactors. Quantum Mechanics found its first technical application in the inferno of the atomic reactor, where streams of neutrons split up the nuclei of heavy atoms and generate heat and electricity. Physicists then turned to the light nuclei, the isotopes of Hydrogen, in attempts to extract more energy from nuclear fusion. For the first time in the history of mankind, mathematical philosophy gave birth to the unprecedented sourced of nuclear power. The discovery of nuclear fission and its utilization in industrial energy production removed any doubt about the immense power of nuclear forces.

Third Stage: High-Energy Elementary Particles
The third period began after World War II with the invention of nuclear reactors, radiospectrometers, particle accelerators, and high altitude flights. This stage could be properly described as the phase of *high energy exploration of the nuclear elementary particles*. Dirac's antimatter postulates initiated the search for new anti-matter particles. During this stage, Quantum Mechanics came up against ever greater difficulties, setbacks and failures. The newly discovered behaviors of elementary particles and their interactions shook the foundation of the old laws upon which Quantum Mechanics was erected.

Since the discovery of the unique spectrum of elements in 1850's, experimental data accumulated at greater rate that challenged many established theories. First, came the law of quanta, the law of conservation of electron motion in specified atomic shells, then the law of conservation of the spin, then the law of conservation of parity, and lastly, the law of conservation of strangeness. The limitations of old Quantum Mechanics are now becoming ever clearer.

The Quantum Theory was founded on the pillars of Classical Mechanics of conservation of total energy and total momentum. It then gained its first breaths after the discovery of the quanta and photons. Quantum Mechanics, thereafter, discovered new conservation laws that resulted from its synthesis with the special theory of relativity.

Today, Quantum Mechanics confronts the following challenges:
1. A rigorous classification of all known and unknown elementary particles.
2. The structure and internal properties of particles of matter.
3. The nature of the forces operating in atomic nuclei.
4. Exact interrelationships between matter and field.
5. The interdependence of mass, time, and space.

CHAPTER 1:
THE ROAD TO THE ATOM: 1792-1900

The Creation From Nothing

Since the beginning of recorded history, instinctual behaviors dominated all living species in such perplexing fashion that defies any logical causation beyond the need for immediate survival. Men from thousands of races and languages worshipped supreme figures despite being unable to come in direct contact with such deity. Such supreme creator was viewed as being able to create live matter from nothing. Hence came the theory of *'creation from nothing'*, which prevailed for many centuries.

The worms of the flies swarming the rotten organic matter were thought to display such exquisite ability of the creator to initiate life from nothing. Even though the discovery of the microscope cracked the immediate theory of creation from nothing (by proving that the worms hatch from eggs lied by the flies in the rotten matter), it would be three centuries before scientists stumbled again on the origin of the DNA. In fact the theory of *'creation from nothing'* is still undisputed as the only mean of assembling atoms to make live DNA. As we will see latter, the supreme deity of the past fictional world is still supreme in our sensual world.

Leonardo Da Vinci (April 15, 1452 -May 2, 1519), Italy.

Galileo Galilei (February 15 1564- January 8 1642), Italy.

The inevitable transition from the past fictional world to the new sensual world was thrust by the need of men to overcome disease and secure resources in order to feed their growing population. During the Renaissance, many great minds engaged in problems that later formed the basis of the new sensual world of Classical Mechanics. Leonardo da Vinci, Galileo Galilei, the Dutch mathematician Simon Stevin and the Frenchman Blaise Pascal are few among those thinkers. Isaac Newton collected the scattered studies of those pioneers and constructed a single unified and harmonious theory for the motion of bodies; terrestrial or galactic.

The Road To Conservation

Among those great minds that paved the road to the establishment of Newtonian Mechanics are the following:

Da Vinci's work signaled the end of the era of the Dark-Ages and the beginning of Renaissance that encompassed a flowering of literature, science, art, religion, and politics, and a resurgence of learning based on classical sources, and gradual but widespread educational reform. In 1482. Da Vinci painted and sculpted. designed buildings. weapons and machinery. and produced many studies on flying machines. nature. geometry. municipal construction. mechanics and architecture. Such **broad interest in establishing knowledge** was a natural result of the years that preceded the industrial age. In those days when Da Vinci lived. there were no fast means of communication beyond horse-drawn wagons and manpowered oar ships. Thus. Da Vinci's work was limited to his immediate locales. Likewise. outside ideas from other states could have not reached thinkers in such rate that allows exponential growth of new ideas.

Da Vinci's Renaissance laid the seeds of the age of Enlightenment that sought to mobilize the power of reason, to reform society and advance knowledge. It promoted science and intellectual interchange and opposed superstition, intolerance and abuses in church and state. In 1610, Galileo was able to verify Copernicus theory that the sun is the center of the cosmos, the earth orbits the sun like other planets, and that other planets have moons like ours. Galileo discovered the Jupiter's four moons and described stars of which only a few can be observed without special means. He rejected the ideas of Ptolemy and Aristotle, which encountered opposition of the Church and scientists. He was even forced to proclaim that his statements of the Earth not being the center of the Universe, were only 'vague assumptions' in stead of facts. Even though Galileo did not acknowledge Kepler's elliptical orbit of the earth around the sun that could explain the varying speed of the earth, he contented himself by winning the battle of getting sun-centered cosmos to replace the earth-centered cosmos. Thus, **Galileo raised great doubts over Aristotelian philosophy,** advanced our knowledge from those broad ideas of Da Vinci into more focused theories about the real architecture of the universe.

Simon Stevin (1548-1620), Belgium.

Blaise Pascal (June 19, 1623 - August 19, 1662), France.

7

As we move from Italy to the west, we observe similar flourishing of objective scientific developments that will soon culminate in Cambridge, England into a new school of classical sciences. On the east shore of the English canal, Simon Stevin lived before Isaac Newton, was a Flemish mathematician and military engineer. Stevin was the first to show how to model regular and semi regular polyhedral by delineating their frames in a plane. He also distinguished stable from unstable equilibria. Stevin discovered the hydrostatic paradox, which states that the downward pressure of a liquid is independent of the shape of the vessel, and depends only on its height and base. He was the first to explain the tides using the attraction of the moon. With such advanced development in civil engineering and extraterrestrial sciences, we could appreciate the prelude to the birth of Newtonian Mechanics.

When one star dies another is born. Pascal was born, in France, two years after Simon Stevin's death in Belgium and shared half of his life with Galileo's in Italy, the another half with Isaac Newton, in England. Pascal made important contributions to the study of fluids, and clarified the concepts of pressure and vacuum. Pascal's law $\Delta P = \rho g(\Delta h)$, or the Principle of transmission of fluid-pressure states that "pressure exerted anywhere in a confined incompressible fluid is transmitted equally in all directions throughout the fluid such that the pressure ratio (initial difference) remains the same. ΔP is the hydrostatic pressure, ρ is the fluid density, g is acceleration due to gravity, Δh is the height of fluid above the point of measurement, putting ρ = mass/volume = m/V, Pascal's law reads $(P_1)(V_1) = (P_2)(V_2)$. which is the fundamental law of gases.

Newton Laws Of Conservation

The wave of Renaissance that started in Italy by Da Vinci a century earlier had arrived to England by mid 1700's with the birth of the age of Enlightenment. In the year 1687, Newton laid the broad foundation of Classical Mechanics in the form of three basic principles that will guide all physical sciences till the present day. The first law of *inertia* explained that objects are unable to change their own state of motion due to their content of matter. The second law of *motion* related any such change in motion of an object to its content of matter as well as to the magnitude of the external or internal force imparted upon the body. The third law of *action and reaction* balanced forces by opposite forces.

The immense significance of Newton's recapping of the accumulated knowledge of two thousand years of civilization into three, simple, succinct, and precise laws lies in the unification of both mythology and philosophy into science. Newton's three laws described the **deeds**, the **doing**, and the **doer** in much fewer words than the Scripture without compromising its meaning.

Isaac Newton (January 4 1643 – March 31 1727), England.

(1) **Inertia** accounted for the need for a doer to cause inert things to move: Mass--> Velocity: Amount of motion = m**v**

(2) **Motion** corresponds to the magnitude of the doing in the amount of motion: Velocity --> Acceleration:
$\mathbf{F} = \partial (m\mathbf{v})/\partial t$

(3) **Action creates reaction,** deeds create deeds, : Acceleration --> Deceleration: $\mathbf{F}_1 = -\mathbf{F}_2$

The Newtonian or Classical Mechanics comprise the governing laws of matter during motion and rest. It applied to physics, chemistry, biology, astronomy, and every physical science that dealt with material objects. The Newtonian Mechanics stood on three pillars: the conservation of mass, the conservation of momentum, and the conservation of energy.

The genius of Newton, beside the simplicity of the three laws, offered a measure for the invariability of motion in terms of the *total energy* and *total momentum* of the moving particles. Those invariable quantities will be used to erect the new science of Quantum Mechanics with the aid of newly discovered laws of conservation, relativity, and quantization of energy.

The broad features of Newton's laws of motion were inevitable outcome of the sparsity of research centers and unstructured educational institutions. It was then clear that man had long been detoured into routes of futile knowledge while the immediate universe shows glaring richness of practical knowledge with every sun rising and sun sitting. Newton contented himself with the undisputed routines of things. Questions such as: **What constitutes matter? What initiates an action of force?** would have to wait three centuries before Quantum Mechanics rose from the ashes of Classical Mechanics and started digging into the conceptual world of mathematics in search for clues to the working of the creator.

Beginning Of The Steam Age

Before marching into the electric age of the nineteenth century, Newton's laws had been honed in harnessing the steam energy which started a new era of *powerful engines*. James Watt worked on the thermal energy of steam and introduced the unit of power: 'watt'. That will later become the major conversion link among all types of energies. The '**Watt**' is used in measuring the power expressed in terms of the force, acceleration, and change of state of motion governed by Newton's laws.

The gradual but progressive advancements in the industrial age are best demonstrated by the evolution of the locomotion system. As early as 1550, roads of rails called Wagonways were being used in Germany. These primitive railed roads consisted of wooden rails over which horse-drawn wagons or carts moved with greater ease than over dirt roads. Wagonways were the beginnings of modern railroads. By 1776, iron had replaced the wood in the rails and wheels on the carts. Wagonways evolved into Tramways and spread though out Europe. Horses still provided all the pulling power. On February 22, 1804, the first steam engine tramway locomotive was commissioned to haul a load of 10 tons of iron, 70 men and five extra wagons the 9 miles between the ironworks in Wales to the bottom of the valley called Abercynnon. It took about two hours.

The contribution of the steam age to modern physics was undoubtedly both direct and indirect. The steam engine put Newton's laws into real life applications and established the conversion units of measurements across all conventional sources of energy that could impart force. The steam age was a prelude to the gas fueled engines and the conventional power stations. Both enabled man to travel long distances in short times and to move greater resources across the continents. The new age of fast communication arrived with trains replacing horses, and steam ships replacing oars. The fast travel of man reflected on the high speed of spread of knowledge and sharing of ideas.

James Watt (19 January 1736 – 25 August 1819), Scotland. Watt made improvements to the steam engine and developed the concept of horsepower.

The French Revolution

The of events, starting from the end of a thousand years of **Dark Ages**, engulfed Europe in a cultural movement of Renaissance that spanned the period roughly from the 14th to the 17th century, beginning in Italy in the Late Middle Ages and later spreading to the rest of Europe, and though the invention of printing sped the dissemination of ideas from the later 15th century. As a cultural movement, it encompassed innovative and gradual but widespread educational reform.and other techniques of rendering a more natural reality in the contributions of such polymaths as Leonardo da Vinci and Michelangelo.

In politics the **Renaissance** contributed the development of the conventions of diplomacy, and in science an increased reliance on observation that would flower later in the Scientific Revolution beginning in the 17th century. Traditionally, this intellectual transformation has resulted in the Renaissance being viewed as a bridge between the Middle Ages and the age of Enlightenment or "The Age of Reason", in 18th-century Europe.

The goal of the **Enlightenment** was to establish an authoritative ethics, aesthetics, and knowledge based on an "enlightened" rationality. The movement's leaders viewed themselves as a courageous, elite body of intellectuals who were leading the world toward progress, out of a long period of irrationality, superstition, and tyranny which began during a historical period they called the Dark Ages. This movement provided a framework for the **American** and **French Revolutions**, as well as the rise of capitalism and the birth of socialism.

The **Industrial Revolution** comprised the direct fruits of Enlightenment from the 18th to the 19th century where major changes in agriculture, manufacturing, mining, transportation, and technology had a profound effect on the social, economic and cultural conditions of the times. It began in the United Kingdom, in 1712, by Thomas Newcomen patents of the atmospheric steam engine and ended in the United States, in 1942, by John Atanasoff and Clifford Berry building the first electronic digital computer and spread throughout Western Europe, North America, Japan, and eventually the world. The **Information Age** followed.

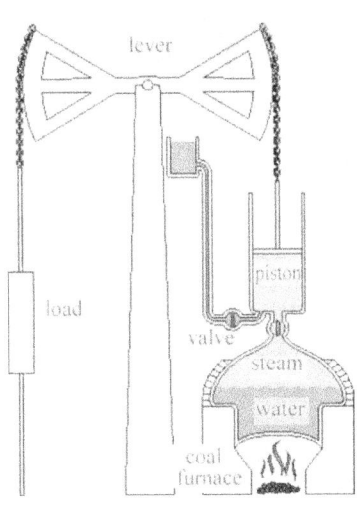

Schematic Newcomen steam engine of 1712, England.

A replica of the Atanasoff-Berry Computer, of 1941, USA, at Durham Center, Iowa State University.

The Industrial Revolution marks a major turning point in history; almost every aspect of daily life was influenced in some way. Most notably, average income and population began to exhibit unprecedented sustained growth. In the two centuries following 1800, the world's average per capita income increased over tenfold, while the world's population increased over six fold.

The role of scientists and research in strengthening the national security of the state was tested and challenged by the **French Revolution** of 1789. Napoleon Bonaparte knew of no boundaries for expansions, invasions and excessive use of force. His Egyptian expedition included a group of 167 scientists: mathematicians, naturalists, chemists and geodesists among them; their discoveries included the Rosetta Stone, and their work was published in the Description de l'Égypte in 1809.

Joseph-Louis Lagrange (25 January 1736 - 10 April 1813), France.

Charles-Augstin de Coulomb (June 14th 1736- August 23 1806), France.

Bonaparte's realization of the immense role of science led to revolutionized European armies and played out on an unprecedented scale. While Bonaparte was recarving the boundaries of European states and for the historian of science in the quiet of the few laboratories that existed in those days, there was in progress a radical reevaluation of the nature of things. Conceptions that had appeared quite stable were being reconsidered. The chemists, physicists and mathematicians of that time made a whole series of outstanding discoveries that prepared the way for the flourishing of the exact sciences in the latter half of the nineteenth century.

In 1793 the Reign of Terror commenced and the Académie des Sciences, along with the other learned societies, was suppressed on 8 August. The weights and measures commission was the only one allowed to continue and Lagrange became its chairman when others such as the chemist **Lavoisier, Borda, Laplace, Coulomb, Brisson and Delambre** were thrown off the commission. In September 1793 a law was passed ordering the arrest of all foreigners born in enemy countries and all their property to be confiscated. Lavoisier intervened on behalf of Lagrange, who certainly fell under the terms of the law, and he was granted an exception. On 8 May 1794, after a trial that lasted less than a day, a revolutionary tribunal condemned Lavoisier, who had saved Lagrange from arrest, and 27 others to death. Lagrange said on the death of Lavoisier, who was guillotined on the afternoon of the day of his trial:-

It took only a moment to cause this head to fall and a hundred years will not suffice to produce its like.

Lagrange was one of the creators of the calculus of variations and invented the method of solving differential equations known as variation of parameters, applied differential calculus to the theory of probabilities. He studied the three-body problem for the Earth, Sun, and Moon (1764) and the movement of Jupiter's satellites (1766). He transformed Newtonian mechanics into a branch of analysis, Lagrangian mechanics. Applied differential calculus equipped **James Clarke Maxwell,** in 1860, with a powerful mathematical method that brought the laws of conservation of electrical and magnetic energies in congruence with Newton's laws. Those same calculus techniques would be used by Schrödinger and Heisenberg, in 1925, in formulating the wave functions of elementary particles.

Coulomb is best known for Coulomb's law describing the electrostatic interaction between electrically charged particles, first published in 1783. The magnitude of the electrostatic force (**F**) between two point electric charges (Q_1 and Q_2) is directly proportional to the product of the magnitudes of each of the charges and inversely proportional to the square of the distance (r) between the two charges. $\mathbf{F} = \varepsilon \, Q_1.Q_2 / r^2$ (ε: the proportionality constant). Coulomb's law is one among the pillars of Quantum Mechanics that establishes the electric forces between charged particles.

Augustin-Jean Fresnel (10 May 1788 - 14 July 1827), France. He was born during the French Revolution and established the wave theory and corpuscular theory of light which were used latter in electromagnetism. Fresnel's wave theory will be used by Maxwell in 1860's in formulating the wave theory of electromagnetism. That theory will latter be modified by Schrödinger in 1925 to accommodate the newly discovered subatomic particles.

Physicists Thomas Young (1773-1829), England, and Fresnel in France laid the foundation of the wave theory of light.

In 1808, Dalton formulated the modern atomic theory, discovered information on gas laws, atomic weight. Dalton's Atomic Theory held:

The atoms of a given element are different from those of any other element; the atoms of different elements can be distinguished from one another by their respective relative atomic weights. All atoms of a given element are identical. Atoms of one element can combine with atoms of other elements to form chemical compounds; a given compound always has the same relative numbers of types of atoms. Atoms cannot be created, divided into smaller particles, nor destroyed in the chemical process; a chemical reaction simply changes the way atoms are grouped together. Elements are made of tiny particles called atoms.

In less than eight years, Dalton's Atomic Theory was given a big blow by Prout's radical hypothesis that atoms of elements heavier than Hydrogen were composites of Hydrogen. This required a century before Prout's hypothesis was fully vindicated, helped the discovery of isotopes, filled the missing spaces in the Periodic Table, and led to the discovery of the neutron in 1932. Dalton's 1808 Atomic Theory was founded on many serious contradictions:

(1) Nature produced **many distinct elements**, in addition to the impossibility of breaking down or building up elements from other elements. That implied that Nature lacked simple, universal laws that could govern the construction of the whole universe from few or single brick of matter or energy.

(2) The **indivisibility** of atoms was based on the little known chemical knowledge of the early 1800's, when no structured relationships between elements were yet known during an astonishingly progressive period of history that unraveled the new laws of electricity.

Prout's hypothesis was an early nineteenth century attempt to explain the existence of the various chemical elements through a hypothesis regarding the internal structure of the atom. In 1815 and 1816, Prout published two papers in which he observed that the atomic weights that had been measured for the elements known at that time appeared to be *whole multiples of the atomic weight of Hydrogen*. He then hypothesized that the **Hydrogen atom** was the only truly fundamental object and that the atoms of other elements were actually groupings of various numbers of Hydrogen atoms.

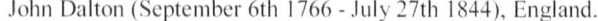

John Dalton (September 6th 1766 - July 27th 1844), England.   William Prout FRS (15 January 1785 – 9 April 1850).

The race between the two rival superpowers; England and France, sprouted the seeds of scientific invention and exploration that reflected upon the exponential rate of growth of modern sciences. Thus, the French Revolution coincided with the most crucial period in the human civilization when the steam age has just begun, the electric age was newly born, and great scientific minds engaged in new research and discoveries. The revolution gave birth to the electrical age, as did the WWII to the age of information.

Birth Of Electricity
Domesticating the newly found electricity into Newton's temple of Classical Mechanics required the earnest efforts of scientists from many states. Initially, electricity was viewed as the flow of something mysterious but that could be confined to volumes, lengths, properties of matter exactly as fluids flow in pipes. Along those concepts, scientists directed their efforts to finding the nature of flow of electricity, long before the electron was discovered in 1895.

Volta introduced the first battery in 1800, made from alternating layers of zinc and copper, which provided scientists with a more reliable source of electrical energy than the electrostatic machines previously used. Volta described the relationships between electrical capacitance (C), electrical potential (V) and charge (Q), and formulated Volta's Law of capacitance. As of yet, none of those quantities was yet convertible to the standards units of mass, length, and time.

Alessandro Giuseppe Antonio Anastasio Volta (February 18, 1745- March 5, 1827), Italy.

Alessandro Volta's electric battery prototype

Hans Christian Ørsted (August 14 1777 - March 9 1851), Denmark.

Michael Faraday (22 September 1791 – 25 August 1867), England.

On July 1820, Ørsted proved that an electric current produces a circular magnetic field as it flows through a wire. The new mystery of electricity that distinguished it from matter and attached field properties to it was discovered by Ørsted.

It would be another forty years before the field properties of electricity blossoms into the wave theory electromagnetism, which opened the doors for wireless telecommunication.

He was to experimental science as was Isaac Newton to theoretical sciences, two towering geniuses of physical sciences. Faraday studied the magnetic field around a conductor carrying a DC electric current and established the basis for the electromagnetic field concept in physics. Faraday discovered electromagnetic induction, diamagnetism, and laws of electrolysis. He established that magnetism could affect rays of light and that there was an underlying relationship between the two phenomena.

Michael Faraday invented the electric motor in 1821. In 1831, using his invention the induction ring, Michael Faraday proved that electricity can be induced (made) by changes in an electromagnetic field which led to the invention of electrical transformers. His inventions of electromagnetic rotary devices formed the foundation of electric motor technology, and it was largely due to his efforts that electricity became viable for use in technology.

André-Marie Ampère (1775-1836), France. He formulated the Ampere's Law in 1826. It became a pillar of the new science of electricity as it relates the integrated magnetic field around a closed loop to the electric current passing through the loop.

Georg Simon Ohm (16 March 1789 – 6 July 1854), Germany, determined the direct proportionality between the potential difference (voltage) applied across a conductor and the resultant electric current, in 1826. This relationship is now known as Ohm's law.

As such, by 1826, Volta, Ørsted, Faraday, Ampere, and Ohm had laid down the governing mathematical laws of the flow of electricity in conductors and succeeded on applying those laws onto the design of practical industrial machines that would soon revolutionize the whole world. Thus, in just two decades, electricity entered Classical Mechanics through the gates of mathematical reasoning and in the form of electrodynamic flow of unseen particles through conducting pipes. Electricity flowed in conductors from high potential to low potential, confronted resistance, and performed work in the same manner that water flowed from high mountains to valleys. The governing laws of the flow of fluids applied perfectly on electrodynamic flow except the new feature of induction. The action of electricity of distant object is now different from Coulomb's 1783 law of repulsion and attraction. Now, Faraday's electric transformers could transfer exact amounts of power from a primary circuit to a secondary circuit without direct conductor and through induced magnetic fields.

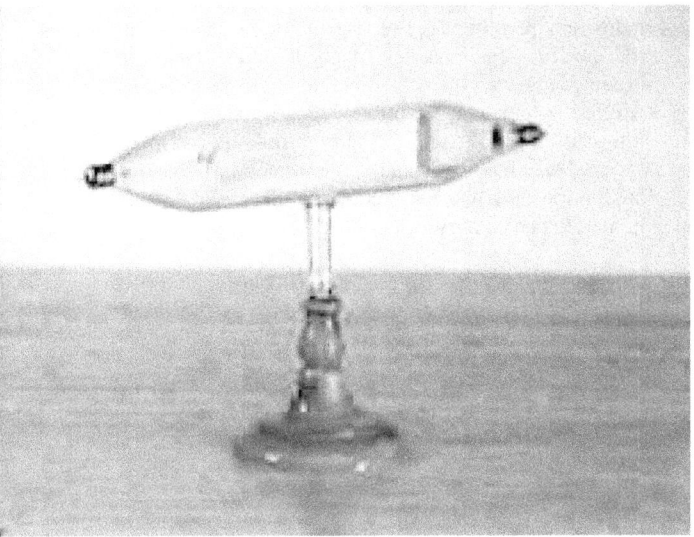

Sir William Crookes (17th of June 1832 -4th of April 1919), England.

The mysterious fluid of electricity started provoking the curiosity of experimentalists. There must be something unusual in electrical conductors that eluded man since the beginning of civilization. For that purpose, Crookes worked on the spectroscopy electrified gases which lead him to his invention of the Crookes tube. The passage of electricity in rarified gases should open a new window on the intrinsic behavior of matter during electrical flow.

William Crookes' tube was an early experiment electrical discharge tube. With his invention cathode rays (electrons) were discovered. Cathode rays are steams of electrons observed in vacuum tubes (Crookes Tube) that are equipped with at least two metal electrodes to which a voltage is applied. Crookes also discovered an unknown element with a bright green emission line in its spectrum. Crookes named this previously unknown element Thallium, from the Greek word which means a green shoot. Also, Crooke is credited to identify the first known sample of Helium in 1895. He also was also the inventor of the Crookes radiometer, which consists of an airtight glass bulb with a partial vacuum.

The new era of experimental exploration into the **subatomic world** started with Crookes' vacuum tubes. In vacuum, the bulk of the colossal numbers of molecules is removed, which allowed the study of pencils of charged particles and analyzing their behavior in the magnetic and electric fields.

Landmarks Of The Nineteenth Century

The Nineteenth Century represents the peak of the industrial revolution. The steam trains and ships transported greater masses of goods across land and oceans, at greater speed than before. The telegram and wireless telecommunication sent and received information across thousands of miles at enormous speed. Thus, man was finally able to conquer the vast land of the earth and communicate at the speed of light in merely one century.

On the relationship of electricity to energy, in 1841, James Prescott Joule showed that energy is conserved in electrical circuits involving current flow, thermal heating, and chemical transformations. A unit of thermal energy, the **Joule**, was named after him. In, 1844, Samuel Morse invented the electric telegraph, a machine that could send messages long distances across wire.

On the relationship of electricity to radiation, in 1860's, Maxwell published the mathematical theory of electromagnetic fields. Maxwell's Equations unified magnetism, electricity and light into four laws of electrodynamics. Like Newton's three laws of motion, Maxwell's Equations are characterized by succinct formulation and ease of application to real-life physical problems. In the three-dimensional vector analysis, those four equations take the following forms:

Alfred Nobel (Born in 1837) founded the most prestigious award for exceptional contributions to science as well as other fields, since the 1900. The annual awarding of Nobel Prize in Physics maintained the steady progress in Quantum Mechanics among many other fields and enticed young and talented scientists to pursue research and invention.

(1) Gauss Law for the Electric Field attributes the divergence of energy through the electric field **E** to the presence of **electric charge** ρ in the volume of space emitting electromagnetic energy. Thus.

$$\text{div } \mathbf{E'} = \rho \qquad \text{or} \qquad \overline{\nabla} \cdot \overline{E} = 4\pi\rho$$

(2) Gauss Law for the Magnetic Field attributes the vanishing divergence of energy through the magnetic field **B** or **H** to the presence of a **dipole field** so no matter how small the volume is you will always find equal numbers of north and south poles, canceling the divergence. Thus

$$\text{div } \mathbf{H} = 0 \qquad \text{or} \qquad \overline{\nabla} \cdot \overline{B} = 0$$

(3) Faraday's Law related the production of magnetic flux to the circuiting of electric current.

$$\text{curl } \mathbf{E} = -\frac{\partial H}{\partial t} \qquad \text{or} \qquad \overline{\nabla} \times \overline{E} = -\frac{1}{c}\frac{\partial \overline{B}}{\partial t}$$

(4) Ampere's Law with Conservation related the production of electric current by the circuiting of magnetic field.

$$\text{curl } \mathbf{H'} = \frac{\partial E'}{\partial t} + j \qquad \text{or} \qquad \overline{\nabla} \times \overline{B} = \frac{4\pi}{c}\overline{J}$$

17

Maxwell's Equations lend great help to the progresses made in wireless transmission of electric power, radios, and television. Through those four wave equations, students could learn the strict meaning of conservation of electric and magnetic energies in terms of the Newton's laws, yet with the newly added *action at a distance* of the electromagnetic fields.

On the <u>transmission electricity across vast lands</u>, in 1883, Nikola Tesla invented the "Tesla coil", a transformer that changes electricity from low voltage to high voltage making it easier to transport over long distances. The transformer was an important part of Tesla's alternating current (AC) system, still used to deliver electricity today.

Now, it was man's turn to <u>travel faster on land</u>. In 1886, automobiles with gasoline-powered internal combustion engines were produced independently by Carl Benz and Gottlieb Daimler. That was soon followed by the invention of wireless communication by Marconi.

James Clerk Maxwell (13 June 1831-5 November 1879), Scotland.

Marconi watching associates raise kite antenna at St. John's, December 1901.

Guglielmo Marconi (25 April 1874- 20 July 1937), Italy. Marconi developed a radio telegraph system that enabled wireless communication across vast distances.

The new age of fast travel and communication thrust mankind into the 20th century with entirely and newly invented technical gadgets that enabled scientists to compete at feverish speeds in innovation and advancement of sciences.

Finally, mounting evidence hinted to the close nature of electricity. In 1897, the Electron was discovered by Joseph John Thomson, which will be discussed in more details in the following chapters.

Ethereal Space

While electricity was already utilized in both direct conductors and wireless communication, its mode of transmission in the void vacuum still defies discovery. Marching into the industrial age was only possible by the invention of the telescope and microscope. Both devices presented man with undisputed evidence of existence the unseen worlds, large and small. That sufficed for the rejection of old beliefs founded on mere imagination and delusion. Yet, the sensual world was still shackled by the limitations of its own senses and instruments. For instance, the **aether** medium that could have explained the transport of energy in vacuum was rejected based on the long standing doctrine that scientific facts must be reproduced and confirmed. That left Classical Mechanics with the unexplained fact that light travels in vacuum, devoid of material content.

Then, **aether** (or ether) was proposed as absolutely solid, elastic, and just as absolutely transparent that it allows for all kinds of bodies moving freely. The failure of experiment to detect the presence of ether as a rootless entity would require three decade explaining the flaws of those experiments. The ether's rootless entity will be substituted by other rootless entities with greater appeal despite their shaky verification by experiment.

In 1932, Dirac redefined the emptiness of space with filled repository of matter and antimatter that go undetected by virtue of their lacking of proper rest energy. Thus, while Dirac's unorthodox conception of the void survived harsh rejection; it never substituted **aether** or explained how fields exist in the absence of a medium.

Even if photons were accepted as particles as Newton and Einstein have proposed, their wave or field nature would still require medium for transmission according to the classical mindset.

The newly invented Relativistic Mechanics, in 1905, did not put the issue of aether to rest, but rather relied on three serious hypothetical facts:

(1) The constancy of the **speed of light** in vacuum;
(2) The **equivalence of matter to energy** through Einstein's equation: $E = mc^2$.
(3) Both matter and field cause **spacetime curvature** equivalent to gravitational field that imparts uniform acceleration.

While Relativistic Mechanics implied that energy content curves spacetime, it did so without rendering mechanism of such effect, for which aether was proposed.

Light Spectrum Of Elements

Inside the atom, greater challenges of no less complexity than aether, mounted, throughout optical experiments of hot gases. Skepticism rose on the classical splitting of the ray of sunlight into a band of many colors, after passing through a glass prism, which was taken as a general law of the refraction of light.

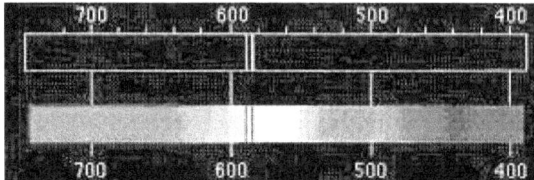

The absorption (top) and emission (bottom) spects of sodium.

Splitting of light spectrum by a glass prism

In 1859, the German chemist Bunsen repeated Newton's old experiment by placing a glass prism in the pathway of the sun's rays and decomposing the light into a spectrum. In Bunsen's experiment, the sun light was substituted by a burning rag dipped in a salt solution. Bunsen didn't see any band at all. When the rag had table salt (**Sodium Chloride**) on it, the spectrum exhibited only a few narrow lines, nothing else. One of the lines was a bright yellow. Bunsen concluded that the role of the glass prism consisted only in sorting the incident rays of light into their wavelengths.

The extended band of the solar spectrum indicated that all the wavelengths of visible light were present. The yellow line, which appeared when the light source was a burning rag, indicated that the spectrum of table salt had a single specific wavelength. When the Sodium was replaced by Hydrogen, in hydrochloric acid, and placed in the flame of a Bunsen burner, the yellow line had disappeared without a trace. That meant that the yellow line belonged to Sodium. This was verified once again. The Sodium was retained, and the Chlorine was replaced in caustic soda, NaOH. The familiar line appeared in the spectrum immediately. There was no longer any doubt. No matter what the substance in which Sodium appeared, it made its whereabouts known by the bright yellow spectral line.

Later, it was found that Sodium is no exception in this respect. **Every chemical element has its own characteristic spectrum.**

As a rule, some of the spectra were much more complicated than that of Sodium and consisted at times of a very large number of lines. But no matter what the compound or substance the element appeared in, its spectrum was always distinct. Spectral analysis did not only help **identify elements in very small traces and beyond any doubt, but it also helped identify elements in inaccessible places such as the distant stars or in the inferno of blast furnaces and in plasma.** Today there are over a hundred chemical elements, and nearly all of them have been classified according to their characteristic spectra.

Discovery Of The Electron

On 30 April 1897, Thomson was the first to propose that the fundamental atomic particles were over 1000 times smaller than an atom, suggesting the sub-atomic particles now known as electrons. He estimated the mass of cathode rays by measuring the heat generated when the rays hit a thermal junction and comparing this with the magnetic deflection of the rays. His experiments suggested not only that cathode rays were over 1000 times lighter than the Hydrogen atom, but also that their mass was the same whatever type of atom they came from. He concluded that the rays were composed of very light, negatively charged particles which were a universal building block of atoms. By comparing the deflection of a beam of cathode rays by electric and magnetic fields he was then able to get more robust measurements of the mass to charge ratio that confirmed his previous estimates.

Thomson's Method
Principle. The cathode rays are deflected by electric and magnetic fields. By subjecting a narrow pencil of cathode rays to electric and magnetic fields acting at right angles to each other, and measuring the deflections of the pencil caused by the two fields, **e/m** can be determined.

Apparatus. The arrangement used by Thomson is sketched as follows.

Thomson discovered the electron and the isotopes of chemical elements, and invented the mass spectrometer.

Sir Joseph John "J. J." Thomson (18 December 1856 – 30 August 1940), England.

The cathode rays produced in a highly evacuated discharge tube and shot out normal to the surface of the cathode C are reduced to a narrow pencil by making them pass through R fine slits in the anode A and in the metal plug B which is electrically connected with A. The cathode ray pencil, traveling in a straight line with a velocity that is accelerated between C and A but remains uniform beyond A, strikes a fluorescent screen SS and produces a small luminous patch at P. On the path of the pencil, electric and magnetic fields are set up as follows: the electric field is produced by maintaining two horizontal plates D and E at a high difference of potential; this field acts in the plane of the diagram and at right angles to the direction of motion of the cathode ray particles. The cathode ray passing through such a field will be deflected in the vertical plane towards that plate which is positively charged. The magnetic field is produced by an electromagnet MM with its direction perpendicular to the plane of the diagram and hence at right angles both to the electric field and the direction of motion of the cathode ray. Such a field will deflect the cathode ray also in the vertical plane, downwards or upwards, according as the field is directed from front to back or back to front. Thus the pencil of cathode rays can be subjected to orthogonal electric and magnetic fields, i.e., acting in directions perpendicular to one another. The lines of force of the two fields are arranged to act over the *same region*, so, that the cathode ray will be under their influence over the *same length* of path.

Theory. Assuming that the cathode ray consists of charged particles with mass, charge and velocity equal to *m, e,* and *v* respectively, let us consider the effects due to the two fields, first acting separately and then simultaneously.

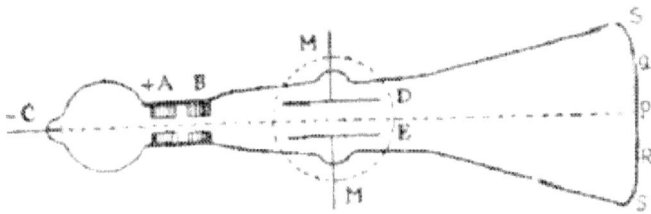

Thomson's apparatus.

Electrostatic deflection. Let the uniform intensity of field between the plates D and E be **X** and the upper plate D positive. Under the influence of this field the cathode ray will be deflected upwards and the luminous patch on the fluorescent screen, which is at P before the introduction of the field, moves up to a position Q. This PQ-shift of the patch can be estimated as follows:

Considering any one particle in the cathode ray pencil, the force exerted on it by the electric field = **X**e (in the direction of the field). Hence the acceleration of the particle in the same direction = **X**e/m. Since the force and acceleration are always constant and perpendicular to the initial direction of the particle, similar to the case of a horizontal projectile subjected to the constant vertical acceleration due to gravity, the path of the particle will be a *parabola*. If *t* be the time of flight of the particle through the electric field of length *l*, the deflection of the particle in the upward direction as it just emerges from the field = $^1/_2(Xe/m)t^2$.
Since *t = l/v*.

The electrostatic deflection = $^1/_2(Xe/m)(l/v)^2$.
Beyond the electric field, as there is no other deflecting agency, the particle moves along a straight line tangential to its path at the point of exit from the field and finally strikes the fluorescent screen at Q. The shift-PQ of the luminous patch depends on two factors, viz., the deflection caused by the field and the distance of the fluorescent screen from the field.

$$PQ = \frac{1}{2} K \left(\frac{Xe}{m} \right) \left(\frac{l}{v} \right)^2 \qquad \ldots (1)$$

where **K** is the constant magnification factor accounting for the geometry of the apparatus.

Magnetic deflection. Let **H** be the uniform intensity of the magnetic field applied at right angles to the plane of the diagram, directed from front to back. The force on the cathode ray particle due to such a field can be calculated as follows:

The force exerted by a magnetic field of intensity **H** on a current element of length *ds* and of strength *i* is **H** *i ds*. Since *i*, the rate of flow of charge = *dq/dt*, the force considered above = **H** *(dq/dt) ds* = **H** *dq(ds/dt)*. In the present case, if *dq* stands for the charge *e* on the particle and *ds/dt* for its velocity *v*, the force exerted on the particle by the magnetic field = **H** *e v*.

21

The direction of the force, as given by Fleming's left-hand rule, will be *vertically downwards,* on account of the direction of the field assumed above. Further, since this force is constant and always acts at right angles to the direction of motion of the particle, the path of the particle in the field is an *arc of a circle.*

If r be the radius of this circular path, we get the relation $\mathbf{H} \, e \, v = m \, v^2 / r$.

The deflection caused by the field may be computed as follows:

Let $\mathbf{y}$ be the deflection of the particle due to its passage in the magnetic field of length l. Since its path in the field is along the arc of a circle of *radius r,* applying the law of segments, $l^2 = y(2r - y) = 2yr - y^2$.

As the deflection y is very small $\mathbf{y}^2$ can be neglected, and $l^2 = 2 \, y \, r$ or $y = l^2 / 2r$.

After emerging from the field the particle moves along a straight line tangential to its circular path at the point of exit from the field, and strikes the fluorescent screen at a point R, vertically below P. The shift-PR of the luminous patch is given by PR = $\mathbf{K} \, y$, where $\mathbf{K}$ is a constant depending on the geometry of the apparatus.

incident cathode ray · deflected ray

$$\therefore \quad PR = K \frac{l^2}{2r}$$

$$\text{But} \quad Hev = \frac{mv^2}{r} \quad \text{and so} \quad r = \frac{mv}{He}$$

$$\therefore \quad PR = K \frac{Hel^2}{2mv} = \frac{1}{2} K \left(\frac{He}{m} \right) \frac{l^2}{v} \qquad \qquad \ldots(2)$$

Simultaneous application of the electric and magnetic fields. Suppose that the separate deflections due to the two fields are in opposite directions. This is obtained by making electric and magnetic fields act at right angles over the same length of path of the ray and suitably arranging the fields, for instance the upper plate D of the electric field being positive and the magnetic field directed from front to back.

Let the intensities of the two fields $\mathbf{X}$ and $\mathbf{H}$ be so adjusted that they produce equal and opposite deflections and in consequence the luminous patch on the SS screen remains in its original position P. Under these conditions, the forces due to the two fields acting on the cathode ray particles just balance each other, so that

$$Xe = Hev \quad \text{or} \quad v = \frac{X}{H} \qquad \qquad \ldots(3)$$

The value of *e/m* of the cathode ray, particle can be determined making use of any one of the two equations (1) and (2) in combination with equation (3).

Experimental procedure. Using the electric field alone of intensity $\mathbf{X}$ the deflection PQ of the luminous spot on the fluorescent screen is measured. The magnetic field also is then switched on and its intensity is adjusted to a value $\mathbf{H}$ which brings back the luminous spot to its initial position P. This is possible only if the two fields are crossed, as explained above.

From equation (1)

$$PQ = \frac{1}{2} K \left(\frac{Xe}{m} \right) \left(\frac{l}{v} \right)^2 = \frac{K}{2} X \left(\frac{e}{m} \right) \frac{l^2}{v^2}$$

and $v = \dfrac{X}{H}$ from relation (3)

$$\therefore \quad PQ = \frac{K}{2} X \left(\frac{e}{m} \right) \frac{l^2 H^2}{X^2}$$

$$\therefore \quad \frac{e}{m} = \frac{2PQ.X}{Kl^2 H^2} \qquad \qquad \ldots(4)$$

In this relation PQ is already measured; **K** the constant depending on the geometry of the apparatus can be evaluated. The intensity of the electric field, **X,** is obtained by dividing the potential difference between the plates D and B by the distance between them; the intensity of the magnetic field, **H,** can be measured by one of the several methods available for measuring the magnetic field between the pole pieces of an electromagnet. The length l of the plate D or E is easily obtained. Hence the value of *e/m* can be determined.

In the *alternate* procedure the magnetic field alone, of intensity **H,** is first applied and the downward deflection PR of the luminous spot caused by it is measured. Next the electric field is made to act along with the magnetic field and the intensity **X** is adjusted to bring back the luminous spot to its initial position P.
From equation (2), we have

$$\frac{e}{m} = \frac{2PR\,v}{Kl^2H}. \qquad \text{But } v = \frac{X}{H}$$

$$\therefore \quad \frac{e}{m} = \frac{2PR\,X}{Kl^2H^2} \qquad\qquad \dots (5)$$

All the quantities on the right-hand side of this equation are experimentally measured and the value of **e/m** can, therefore, be evaluated.
Note. The velocity v can be measured by the simultaneous application of both the electric and magnetic fields so that the deflection caused by one of them is annulled by the other and using relation (3).

Other methods of finding v have also been used. Thus, Thomson deflected the cathode ray pencil by means of a powerful magnet into a small insulated metallic cup containing one junction of a thermocouple and connected to a quadrant electrometer. He measured the total charge conveyed to the cup by the cathode ray particles as well as the heat developed by their stoppage. If **N** particles enter the cup per second,
then the total charge $\mathbf{Q} = \mathbf{N}e$;
and if **W** is the total energy received by the cup per second,

$$W = \frac{1}{2}\,mv^2 N.$$

Hence,

$$v^2 = \frac{2W}{mN} = \frac{2We}{Qm}.$$

Hence, Kaufmann calculated v from the potential difference of the discharge tube as follows:
The work when charge **e** *falls* through a potential difference V is equal to eV.
But the kinetic energy of a cathode ray particle *is*

$$\frac{1}{2}\,mv^2 = eV \qquad\qquad v^2 = \frac{2eV}{m}$$

Results

(1) The value of v obtained by these methods varies between 2.2×10^9 and 3.6×10^9 centimeters/second, a value which is about 1/10 of the velocity of light, hence very high.

With such velocities electrons can reach the moon in a few seconds. The velocity of the cathode ray particles evidently depends upon a number of conditions, chiefly on the gas pressure and potential difference of the discharge tube. If the particles have different velocities a dispersion effect is observed in both the electric and magnetic deflections;

(2) The value of *e/m* is found to be a constant independent of the material of the cathode as well as of the kind of gas in the discharge tube, which, therefore, indicates that the particles in the cathode ray are of the same kind, though they might acquire widely different velocities.

(3) The numerical value of *e/m* obtained by Thomson in his first experiments was of the order of 10^7 *e.m.u.;* but a later determination gave 1.7 x 10^7 *e.m.u.* or 5.1 x 10^{17} *e.s.u.* since the *e.s.u.* is obtained by multiplying the *e.m.u.* by 3 x 10^{10} which is the velocity of light. This value comes very close to that of modern accurate measurements involving entirely different experimental techniques, such as the comparative study of the spectra of hydrogen and ionized helium, the Zeeman effect, Dunnington's method of subjecting thermionic electrons to the combined influence of a high frequency oscillator and a large uniform magnetic field and Beardon's measurement of the refractive index of a diamond prism for x-rays of known wavelengths. The mean value of *e/m* given by these high-precision methods is 1.76 x 10^7 *e.m.u. (or* 1.758820150(44)×10^{11} C/kg). We may, therefore, conclude that the cathode ray particles are identical with electrons that enter into the constitution of all matter.

(4) Comparing the value of *e/m* of the electron with that of *e/M* of the hydrogen ion in electrolysis, which is 9.6494 x10^3 *e.m.u.* (since 1 gram atom of hydrogen. carries 96494 coulombs or 9649.4 *e.m.u.* of electricity) it is seen that the first is about 1840 times as large as the second. This might mean that

> either the charge on the electron is 1840 times as large as the charge on the hydrogen ion, the mass being the same in the two cases; or the mass *m* of the electron is 1/1840 of the mass *M* of the hydrogen ion, the charge on each being the same.

To decide this point it is necessary to determine the charge **e** on the electron separately.

Robert A. Millikan (March 22, 1868- December 19, 1953), USA.   Millikan's Experiment

Millikan accurately determined the charge carried by an electron, using the "falling-drop method"; he also proved that this quantity was a constant for all electrons (1910), thus demonstrating the atomic structure of electricity. The discovery of his law of motion of a particle falling towards the earth after entering the earth's atmosphere, together with his other investigations on electrical phenomena, ultimately led him to his significant studies of cosmic radiation (particularly with ionization chambers).

Thomson Model Of The Atom

In 1898, Thomson believed that the electrons emerged from the atoms of the trace gas inside his cathode ray tubes. He thus concluded that atoms were divisible, and that the electrons were their building blocks. Thomson described atoms as clouds of positive charge within which floated negative electrons in quantities sufficient to balance the charge. The electrons were attracted by the positive clouds and retarded in their motion. To explain the overall neutral charge of the atom, he proposed that the electrons were distributed in a uniform sea of positive charge; this was the **plum pudding model** as the electrons were embedded in the positive charge like plums in a plum pudding. In Thomson's model, electrons were not stationary but were orbiting rapidly. The nucleus center of the atom was not yet perceived and so was the proton.

Thompson's atom model must explain the emitted light spectrum of elements. The source of spectral light was strongly tied with the theory of thermal radiation and bore all the traces of the basic shortcoming of this theory. The basic weakness lies in understanding the mechanism and source of light within the atom.

The old view was that by increasing temperatures, molecules move faster, collide more violently and more frequently, and the molecules vibrate so fast that they begin to emit light. But then why do not bodies luminesce at room temperature, since the molecules are still in motion? According to Classical Mechanics, charged particles have to emit electromagnetic radiation when they are decelerated. As deceleration reduced the kinetic energy of a particles, that energy loss appears as radiation. Apparently, that radiation is the light emitted when bodies are heated.

At first glance, the explanation was quite convincing. The more a body is heated, the faster the electrons move in the atoms and the greater the deceleration due to the attraction of the clouds of positive charge, and hence the more intense the radiation. That could be the case if electrons did not expend energy when radiating. But when electrons radiate light, they must decelerate with extreme rapidity. In just a minute fraction of a second they would have bogged down in the positive clouds like raisins in pudding.

Something was wrong. Several years later it became evident that the Thomson model of the atoms would not work in other respects as well. Too many questions remained unanswered. And then why don't the electrons simply merge with the positive cloud and neutralize their charge? The few answers that are obtainable from this model in most cases come into sharp conflict with experiment.

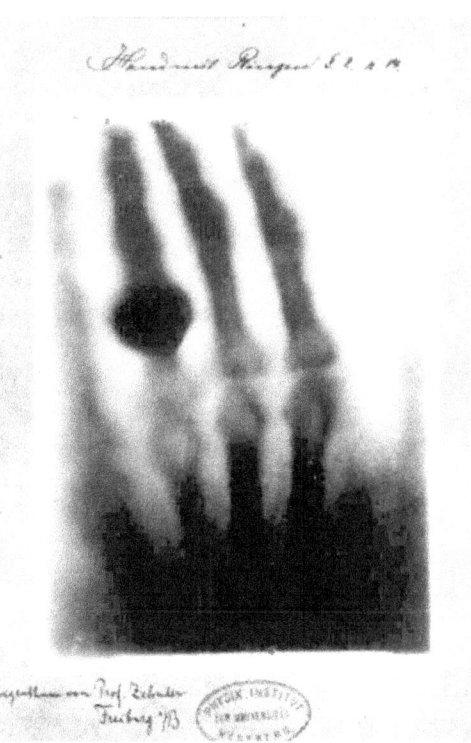

Wilhelm Roentgen (27 March 1845- 10 February 1923), Germany.

First medical X-ray in history, dated 22 December 1895, by Wilhelm Röntgen (1845–1923) of the left hand of his wife Anna Bertha Ludwig. It was presented to Professor Ludwig Zehnder of the Physik Institut, University of Freiburg, on 1 January 1896.

Discovery Of X-Rays

Roentgen discovered a new form of radiation in 1895. He called it X-radiation. On the evening of November 8, 1895, he found that, if the discharge tube is enclosed in a sealed, thick black carton to exclude all light, and if he worked in a dark room, a paper plate covered on one side with barium platinocyanide placed in the path of the rays became fluorescent even when it was as far as two meters from the discharge tube. During subsequent experiments he found that objects of different thicknesses interposed in the path of the rays showed variable transparency to them when recorded on a photographic plate. When he immobilized for some moments the hand of his wife in the path of the rays over a photographic plate, he observed after development of the plate an image of his wife's hand which showed the shadows thrown by the bones of her hand and that of a ring she was wearing, surrounded by the penumbra of the flesh, which was more permeable to the rays and therefore threw a fainter shadow. This was the first "röntgenogram" ever taken. In further experiments, Röntgen showed that the new rays are produced by the impact of cathode rays on a material object. Because their nature was then unknown, he gave them the name X-rays.

Discovery Of Radioactivity

As the playful experimentation with gases, wires, magnets, and radiation probed the nature of the newly discovered electrical energy, an unexpected surprise occurred that threw nuclear energy into the fray, at the dawn of the 20th century. In 1900, science confronted those greatest inquiries into the essence of energy:

(1) **Aether** required verification and detection.
(2) **Optical spectra** of atoms imposed discrete boundaries on waves and particles.
(3) **X-ray** emission confirmed the skepticism of the classical nature of atomic structure.
(4) **Electron discovery** raised question on the balance of charges and mass distribution inside the atom.
(5) **Natural radioactivity** hinted to the perpetual source of energy inside the atom.

Antoine Henri Becquerel (15 December 1852 – 25 August 1908), France, discovered radioactivity along with Marie and Pierre Curie.

Marie Curie (November 7, 1867- July 9, 1934), France, investigated the minerals containing Uranium that gave off rays. Uranium rays were discovered by Becquerel.

Pierre Curie (15 May 1859-19 April 1906) Paris, France

Röntgen's X-rays raised doubts on the material contents of mass since the energetic rays were able to penetrate solids as if they were empty space. On the other had, the naturally emitted radiation, discovered by Becquerel, raised greater

doubts on the classical conservation of energy since radioactive materials emitted energetic radiations without being acted upon by any external forces.

No single canon in Classical Mechanics for the atom was able to account for the mysterious release of energy by Uranium, Radium and other chemical elements-a radiation of energy that continues without interruption for many thousands and millions of years without any outside source.

Classical Senses And Canons

The new discoveries of the subatomic phenomena led scientists to reexamine the axioms of Classical Mechanics in hope for pinpointing the flaws of the old theory. Classical laws grew out of our sensual perception of objects. The whirling of a ball along a stretched rope would not start whirling unless pushed by hand. The greater the magnitude of the push, the faster the whirling and the more stretched the rope.

Even in the absence of a rope, the *action at distance* is observed when a ball keeps rolling along a smooth horizontal table after the action of the hand that pushed the ball has ceased. This observation gave rise to Newton's law of inertia. The law stated the obvious. That objects of material nature are unable to initiate motion or to end it unless aided by an external action; i.e., material objects are inert.

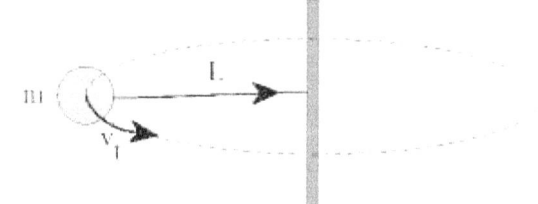

Balance of centrifugal force and central attraction.

The response of the *inert* objects to the external action was described by Newton's second law of motion. It related the magnitude of the action to both the material content of the inert object and to the resultant acceleration of the object.

Ingeniously, Newton extended those sensual laws to the motions of celestial masses such as planets and stars. There is of course no way to perceive directly the action of the force governing a grandiose phenomenon of nature as planetary motions. But the force is there. And Newton discovered it. We know that it is the force of the reciprocal attraction of bodies. Newton's genius lies in the fact that he perceived what is common between the motion of a ball and the orbital motion of a planet.

All along, Classical Mechanics formulated its laws out of observation without the need to explain the underlying causes of actions. It sufficed to assume that objects were all *inert* and actors were all *unknown*.

Birth Of Quantum Mechanics

The nineteenth century witnessed the rise of thermodynamics; optics; and electrodynamics. For a time, physics remained in a rather contented state. All new discoveries continued to fit neatly into the existing moulds. However, as the field of Classical Mechanics grew upwards, its enormous front gave signs of fatigue, sinister cracks appeared, and finally the entire structure began to crumble under the bombardment of new facts.

The first invisible subatomic particle; the electron, has already been discovered in 1897 despite being in use in many industrial applications since the dawn of the nineteenth century. Two decades earlier, Maxwell had formulated the governing wave equations of the fields associated with the motion of the electrons.

Classical Mechanics was entirely satisfactory as long as physics was confined to sizable objects and slow events. That was gradually and progressively challenged by the discovery of *electricity*, followed by *electromagnetic* wireless transmission, then the accidental discovery of *radioactive* materials. The three discoveries deviated significantly from the sensual universe of Newton, towards the unseen universe of subatomic processes:

1. *Electricity* → particles and energy flow in metallic wires causing sparks, light, heat, and magnetic effects.
2. *Electromagnetic waves* → effects of electricity were detected in metallic wires placed remotely from the original source of electricity.
3. *Radioactive materials* → particles and energy flow out of natural materials without man-made source of electricity.

Newton's laws never accounted for the fields associated with matter or with the endless release of energy from radioactive materials without forces being imparted upon them. Further, Maxwell's formulation of the electromagnetism inherited the flaws of Newton's laws in the form of describing the energy of electromagnetic fields as a *continuum*, without discrete boundaries. The new discovery of the spectrometer created a severe blow to the Classical Theory of Electromagnetism. It showed *discrete spectral lines* of atomic radiations rather than a *continuum spectrum*.

Further, subatomic particles lose their dimensions and acquire the properties of waves; waves, in turn, begin to act like particles. Electrons and the other building blocks of matter pass through insuperable barriers or vanish outright, leaving photons in their place.

The stampede towards the subatomic universe advanced on two fronts. The experimental front revealed new discoveries at unprecedented rates and new inventions were catching up with the new discoveries. On the theoretical front, gradual but persistent deviation from the sensual world was in earnest. The theoretical front undulated between lagging behind the new discoveries and leading into newer ones.

The final straw that held Classical Mechanics afloat was drowned by *discrete spectral lines* emitted from heated elements. Both the wave theory of magnetism and laws of conservation failed to account for the discreteness of the energies emitted from heated matter or the spontaneous emission of radiation. Boltzmann's statistical model of atomic transitions would lead the way into the new era of Quantum Mechanics.

Thermal Radiation

The invisible radiation emitted from hot matter is called *heat* or *infrared radiation*. Thermal radiation is quite a common thing in nature. The emission of light and heat is actually one process that occurs even in the stars and governed by well elaborated laws.

First, the more a body is heated, the brighter it glows. The quantity of radiation emitted per second varies drastically with change of *temperature* of the body. If the temperature is increased three times, the radiation will increase almost one hundredfold.

Second, the *color* of the emission changes with an increase in temperature, from dark, then a faint crimson tinge appears, then red, then orange and yellow. And finally the heated body begins to emit a white light.

A black body absorbs the most radiation and, hence, is heated by this radiation to a higher temperature than all other bodies.

Conversely, when a black body is heated to a high temperature and becomes a source of light, it radiates more intensely at the given temperature than any other bodies. This, then, is a very convenient radiator for establishing the quantitative laws of thermal radiation.

However, it was found that black bodies themselves emit radiation in different ways. For example, soot may be blacker or lighter than black velvet, depending on the fuel it comes from. And velvet too can differ.

Black Body Radiation
The need for standardizing measurement of emitted and absorbed radiation was crucial to the establishment of rigorous laws of statistical mechanics. Those laws have already been founded on intricate probability axioms that attached greater uncertainties to the measurement. Thus a measuring device of the quantity of radiation must account for all possible portions of energy being measured, otherwise, uncertainties would turn into chaotic errors. The black surface is the best candidate for the measurement of radiation. Enclosing those black surfaces within a box with a tiny aperture that permits the entry of the rays of radiation to be measures, the black body box will capture the rays of radiation, never let them out again, caught for all time. The black box absorbs all the radiant energy that enters it. Such device is called *"radiometer"*.

Reversely, the black box could be made a source of light when heated sufficiently. The walls become incandescent and begin to emit visible light. As we have already said, for a given temperature the thermal and light radiation of such a box will be greater than for any other bodies, which are then called gray to distinguish them from our box.

The black box presented standard body, a criterion to be used as a basis for establishing the laws of radiation of heated bodies. Then the emission of light by other bodies could be regarded as deviations from this *'standard'*.

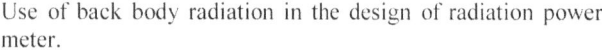

Use of back body radiation in the design of radiation power meter.

Ludwig Eduard Boltzmann (February 20, 1844 – September 5, 1906), Austria.

Boltzmann-Wien Laws

Boltzmann worked in the field of statistical mechanics and statistical thermodynamics that paved the road to the discovery of the Quanta. Boltzmann's equation describes the atomic number distribution in terms of the energies of each group of atoms. Classical statistics is rightly called *Maxwell-Boltzmann statistics*, since it takes its origin from Maxwell's law of distribution of molecular velocities and Boltzmann's theorem relating entropy and probability. Hence it is necessary to review briefly these two basic ideas.

Maxwellian distribution of molecular velocities

According to the kinetic theory of matter, a gas consists of a very large number of rigid and perfectly elastic material particles, called molecules, all of which have the same mass and move freely within the vessel containing the gas. Moving in all possible directions and colliding with each other these molecules acquire all possible velocities ranging from $-\infty$ to $+\infty$. Maxwell in 1859 conceived the idea that, at a given temperature when the gas is in a state of thermal equilibrium, there should be a law according to which the velocities of the molecules can be grouped, in spite of their apparently random and chaotic movements. He was able to derive that law of distribution of molecular velocities by the application of elementary ideas of probability as follows:-

Considering, for the sake of simplicity, a perfect, monatomic gas (where translational motion alone of the molecules is involved, rotational and vibrational motions being excluded) and assuming that the molecular density is unaffected by the molecular motions and collisions, probability considerations show that the number of molecules n_1 which have component velocities lying between

u and $\underline{u} + du$, v and $v + dv$ and w and $w + dw$,

along the three axes of the Cartesian coordinate system is given by

$$n_1 = n\, f\,(u,\, v,\, w)\ dv\ dv\ dw$$

where n is the number of molecules in unit volume. This is Maxwell's law of distribution of molecular velocities, which is usually expressed in the following two forms:-

$$(1) \qquad n_1 = n\, a\, e^{-bmc^2}\ d\tau$$

where

$$a = (b^3 m^3 / \pi^3)^{1/2}$$

$$c^2 = u^2 + v^2 + w^2 \ \text{ and}$$

$$d\tau = du\ dv\ dw.$$

b is a constant,
m the molecular mass,

$$(2) \qquad n_1 = n\ \left(\frac{m}{2\pi k\mathrm{T}}\right)^{3/2} e^{-mc^2/2k\mathrm{T}}\ d\tau$$

where k is Boltzmann's constant and T the absolute temperature of the gas. The identity of (1) and (2) is readily seen by putting $b = 1/2k\mathrm{T}$

Boltzmann's theorem on entropy and probability

A gas at a definite temperature and pressure is in a sensibly constant macroscopic state, but its constituent molecules are in incessant motion of a random and chaotic nature, so that the microscopic state of the gas is continually changing. The reason for this state of affairs must be naturally sought in the fact that

of all the possible microscopic states the vast majority correspond to values of the quantities which make the macroscopic state practically constant.

If any microscopic state has a value very much different from that of the macroscopic state, the probability of its remaining in that state is small. When a substance is not in a state of macroscopic equilibrium, its state changes until equilibrium is established. This process involves naturally an increase in the number of possible microscopic states. Thus, it is evident that an *equilibrium macroscopic state is one for which the number of microscopic states is a maximum.*

On the other hand, according to the second law of thermodynamics in the form proposed by Cornot, the entropy of a system tends always towards a maximum and

the maximum entropy corresponds to the maximum disorder, hence to the statistical condition of maximum probability.

From reasoning like this, Boltzmann concluded that there must be a relation between the thermodynamical entropy which has always a maximum value in cases of equilibrium and the maximum probability of the dynamical equilibrium state. If S is the entropy of an isolated system and W the number of possible microscopic states through which the system passes in a given macroscopic state, both S and W tend to increase to maximum values and according to Boltzmann's idea, S is a function of W:

$$S = f(w)$$

where W may be called the thermodynamic probability of the state of the system. In order to render this relation more explicit, let us consider two separate systems, having entropies S_1 and S_2 and thermodynamic probabilities W_1 and W_2. Then

$$S_1 = f(W_1)$$
and
$$S_2 = f(W_2)$$

The total entropy of the two systems is

$$S_1 + S_2 = f(W_1) + f(W_2) \quad \dots\dots\dots\dots (1)$$

But

the thermodynamic probability of the two systems taken together is $W_1 W_2$. Hence

$$f(W_1 W_2) = S_1 + S_2 = f(W_1) + f(W_2) \quad \dots\dots\dots\dots (2)$$

If this relation is to be satisfied, $f(W)$ must be a logarithmic function of W.

Therefore,

$$f(W) = k \log W$$
and
$$S = k \log W \quad \dots\dots\dots\dots\dots\dots\dots\dots (3)$$

This relation (3) can be established more rigorously as follows:

Considering unit volume of a perfect monatomic gas in thermal equilibrium, let there be n molecules in it. The velocity components u, v, w of each molecule may be represented by a velocity point. Dividing the unit volume into a very large number of elementary volumes assumed equal, these will contain n_1, n_2, n_3,......velocity points, such that

$$n_1 + n_2 + n_3 + \dots\dots = n$$

When the n_1 velocity points, contained in the first elementary volume are permuted among themselves, $n_1!$ complexions are obtained which are indistinguishable from one another, since they correspond to the same macroscopic state of the gas. Similarly, by permuting n_2 velocity points in the second volume element, $n_2!$ equivalent complexions are obtained and so on..

If, on the other hand, all the n velocity points are permuted in all possible ways, $n!$ distributions are obtained. These contain many equivalent complexions and we are interested only in those which remain distinct from one another for defining the macroscopic state of the gas. Now, the number of distinguishable complexions is given by

$$W = \frac{n!}{n_1! \, n_2! \, n_3! \ldots \ldots}$$

On the assumption that n is very large, using Stirling's theorem,

$$\log W = n \log n - n - (n_1 \log n_1 - n_1 + n_2 \log n_2 - n_2 + n_3 \log n_3 - n_3 + \ldots)$$

$$= n \log n - n - (n_1 \log n_1 + n_2 \log n_2 + n_3 \log n_3 + \ldots) + (n_1 + n_2 + n_3 + \ldots)$$

$$= n \log n - (n_1 \log n_1 + n_2 \log n_2 + n_3 \log n_3 + \ldots)$$

$$= - (n_1 \log n_1 + n_2 \log n_2 + n_3 \log n_3 + \ldots)$$

$$= - \Sigma n_1 \log n_1$$

omitting the constant term ($n \log n$), as it does not effect the present consideration.

Replacing the summation by integration and using the relation

$$n_1 = n \, a \, e^{-bmc^2} \, d\tau,$$

We get

$$\log W = - \int n \, a \, e^{-bmc^2} \, d\tau \log (n \, a \, e^{-bmc^2} \, d\tau)$$

$$= - \int n \, a \, e^{-bmc^2} \, d\tau \, (\log n \, a + \log d\tau - bmc^2)$$

$$= - \int n \, a \, e^{-bmc^2} \, d\tau \log n \, a + bmc^2 \int n \, a \, e^{-bmc^2} d\tau$$

since ($\log d\tau$) can he omitted.

$$\log W = - n \log na + bmn \int c^2 \, a \, e^{-bmc^2} \, d\tau$$

And since

$$\int c^2 \, a \, e^{-bmc^2} \, d\tau = \overline{C}^2$$

Where $\overline{C}$ is the mean velocity.

$$\log W = - n \log (na) + bmn \, \overline{C}^2$$

Since

$$\overline{C^2} = 3/2bm$$

Then

$$\log W = -\, n \log (na) + (3/2)\, n$$

Further, as $n \propto 1/V$ and $a \propto b^{3/2} \propto 1/T^{3/2}$, we get

$$\log W = \frac{N}{V} \log (T^{3/2}\, V) + \frac{G}{V}$$

where N is the Avogadro number and G a constant.

This result refers to unit volume. If V is the gram molecular volume,

$$\log W = N \log (T^{3/2}\, V) \qquad \dots (4)$$

neglecting G.

To relate this result with the entropy, let us derive an expression for the entropy S of a perfect monatomic gas in terms of the temperature and the molecular volume.

Let dQ be the heat supplied to the gas which is utilized in doing external work $p\, dV$ and in raising the temperature by dT. Then

$$dQ = C_v\, dT + p\, dV$$

If dS is the change of entropy,

$$dS = dQ/T,$$

so that

$$TdS = C_v\, dT + p\, dV$$

Since $pV = RT$,

$$TdS = C_v\, dT + \frac{RT}{V}\, dV$$

Dividing throughout by T,

$$dS = C_v \cdot \frac{dT}{T} + R \cdot \frac{dV}{V}$$

Integrating,

$$S = C_v \log T + R \log V + \text{a constant.}$$

Omitting the additive constant,

$$S = C_v \log T + R \log V$$

Since, for a monatomic gas $C_v = (3/2)\, R$,

$$\begin{aligned} S &= (3/2)\, R \log T + R \log V \\ &= R\, (\log T^{3/2} + \log V) \\ &= R \log (T^{3/2}\, V) \qquad \dots (5) \end{aligned}$$

Comparing this with relation (4), we get

$$S = (R/N) \log W = k \log W$$

Thus Boltzmann's theorem connecting entropy and probability is proved.

33

In classical thermodynamics the entropy is usually taken as the *difference* between the entropy in the actual state and the entropy in an arbitrarily chosen standard state. If the thermodynamic probability in the standard state is W_o, then

$$S = k \log W - k \log W_0 = k \log (W/W_0)$$

Further, the standard state is referred to that of absolute zero, since, according to Nernst's heat theorem, the entropy at absolute zero is zero so that W/W_o for that state is equal to unity. This means that

the standard state of absolute zero is one of perfect order, in which the molecules are either at rest or move with uniform velocity in parallel rows.

On this basis, Boltzmann's theorem of entropy and probability is written as

$S = k \log N$

where $N = W/W_o$ represents the number of probable ways in which a particular state (other than that of absolute zero) can be realized.

An expression for N may be derived by the use of statistical laws as follows:-

Let us suppose that the entire volume containing n molecules of a gas is divided into a number of separate elementary volumes or *cells,* in which the molecules are distributed. Let these cells be designed by $A_1, A_2, A_3, \ldots\ldots A_s$ and let them contain. $n_1, n_2, n_3, \ldots\ldots n_s$ molecules, each with energy $E_1, E_2, E_3, \ldots\ldots E_s$ respectively, subject, of course, to the following two conditions:

Total number of molecules: $n_1 + n_2 + n_3 + \ldots\ldots n_s = n$

Total energy: $\qquad\qquad n_1E_1 + n_2E_2 + n_3E_3 + \ldots\ldots n_sE_s = E$

The possible number of ways this distribution can be realized is

$$\frac{n\,!}{n_1\,!\;n_2\,!\;n_3\,!\ldots\ldots n_s\,!} = \frac{n\,!}{\displaystyle\prod_{k=1}^{k=c} n_k\,!}$$

To get the probability of this distribution we have still to consider the a *priori* probability of the distribution. For every one of the above possible ways there is no restriction on any one of the molecules occupying any one of the cells, since the molecules are considered *distinct* from one another, (each having a *recognizable individuality* of its own) and each cell can accommodate any number of molecules. Hence, considering the distribution of the n_s molecules in the A_s cells, any one of the n_s molecules can occupy any one of the A_s cells. This means that the probable number of ways in which the n_s molecules can be distributed in the A_s cells is $(A_s)^{n_s}$. In a similar manner, the molecules in the other cells can be grouped as $(A_1)^{n_1}$, $(A_2)^{n_2}$, $(A_3)^{n_3}$, etc. All such groupings can therefore be done in

$$\prod_{k=1}^{k=c} (A_k)^{n_k}$$

ways.

Hence the probable number of ways W, in which the desired distribution can be effected, is given by

$$W = \frac{n\,!}{\displaystyle\prod_{k=1}^{k=c} n_k\,!} \times \prod_{k=1}^{k=c} (A_k)^{n_k} = n\,! \prod_{k=1}^{k=c} \left(\frac{A_k^{n_k}}{n_k\,!}\right)$$

Now, since $W_0 = n!$ is the probable number of ways in which the molecules can be arranged in the state of perfect order, therefore

$$\therefore \quad P = \frac{W}{W_0} = \prod_{k=1}^{k=s} \left(\frac{A_k^{n_k}}{n_k!} \right) \quad \dots (6)$$

Equilibrium conditions and distribution law

The distribution law corresponding to the equilibrium state is governed by the *primary* condition that the state must possess *maximum* probability, as well as by the *two subsidiary* conditions concerning the total number of molecules and the total energy, stated above. For a gas in the steady state, *i.e.,* in thermal equilibrium,

$$S = k \log P = k \log \prod_{k=1}^{k=s} \left(\frac{A_k^{n_k}}{n_k!} \right)$$

$$\therefore \quad S/k = \Sigma \left[n_s \log A_s - \log n_s! \right]$$

By Stirling's theorem,

$$\log n_s! = n_s \log n_s - n_s$$

Hence

$$S/k = \Sigma \left[n_s \log A_s - n_s \log n_s + n_s \right]$$

Since, in the steady state S is constant, $dS = 0$ and $dS/k = 0$.

$$\therefore \quad \Sigma \, dn_s (\log A_s - \log n_s) - \Sigma n_s \, dn_s/n_s + \Sigma \, dn_s = 0.$$

Since $n = \Sigma n_s$ is a constant, $\Sigma \, dn_s = 0$ and also, $E = \Sigma E_s n_s =$ constant, $\Sigma E \, dn_s = 0$. Under these conditions, remembering also that A_s is not subject to variation, the above relation reduces to

$$\log A_s - \log n_s = 0$$

Using the method of *undetermined multipliers,* we obtain

$$\log A_s - \log n_s - \lambda - \beta E_s = 0$$

where λ and β are two undetermined constants, known as *Lagrangian multipliers.*

The above relation can be put in a more convenient form:

$$\log \frac{A_s}{n_s f} = \beta E_s \qquad \text{or} \qquad \frac{A_s}{n_s} = f \, e^{\beta E_s}$$

where β is a constant affecting the energy content of the n_s molecules distributed in the A_s cells and f is an unknown function whose value depends on the conditions of the case and is ordinarily called the *degeneracy parameter.*

The classical distribution law is therefore given by

$$n_s = (A_s/f) \, e^{-\beta E_s} \quad \dots (7)$$

From this relation, we see that

the number n_s corresponding to the cell A_s essentially involves the energy E_s belonging to this cell as well as the size A_s of the cell, and that in such a way that among cells of equal size, one with greater energy is not so well filled as one with smaller energy.

The fall in the value of n_s with increasing energy obeys an exponential law.

35

Maxwell's law of distribution of velocity can be deduced from the above relation by substituting the values of A_s, f and E_s, proper to the case; it can he shown also that $\beta = 1/kT$.

$$\frac{N_b}{N_a} = \left(\frac{g_b}{g_a}\right)\left(e^{-(E_b - E_a)/kT}\right) \quad \text{or} \quad E_b - E_a = kT\ln\left[\frac{g_b}{g_a}\right] - kT\ln\left[\frac{N_b}{N_a}\right]$$

In the study of black body radiation a special interest was attached to the spectral distribution of energy in the emitted radiation.

How was the energy distributed among the various wavelengths in the spectrum and at what wavelength was most of the energy emitted ?

Naturally, the electromagnetic theory, then prevalent, was applied to the problem but it led to wrong and inconsistent results not agreeing with experimental data.

In 1884 Stefan and Boltzmann, using the idea of the pressure exerted by radiation according to the electromagnetic theory and thermodynamical principles, showed that

the total energy density, *i.e.,* the energy of radiation in unit volume of space due to all the different wavelengths in the spectrum was proportional to the fourth power of the absolute temperature of the black body.

This is what is known as *Stefan's fourth-power law* of black body radiation, which evidently does not throw any light on the energy distribution among the different wavelengths in the thermal spectrum but refers only to a globular effect.

In 1893 Wien, in order to find the actual distribution of energy in the thermal spectrum, considered the case of an enclosure full of black body radiation expanding *adiabatically* with a velocity small compared with that of light and proved by thermodynamical reasoning that after such an expansion the radiation still remained full but was characteristic of the new temperature. From a knowledge of the work done by the radiation pressure during the adiabatic expansion, the new distribution of energy could be obtained. In this way he was able to establish two laws known as *Wien's displacement laws, viz.,*

λT = a constant and
ET^{-5} = a constant,

where λ is the wavelength corresponding to the absolute temperature T and E the emissive power. The first law shows an inverse variation between the wavelength and absolute temperature of the source, while the second, a direct variation of the emissive power for a radiation of wavelength λ with the fifth power of the absolute temperature. These two laws can be combined into one general relation which is expressed as

$E = K \lambda^{-5} f(\lambda T)$,

where K is an absolute constant, $f(\lambda T)$ some function of (λT).

Wilhelm Carl Werner Otto Fritz Franz Wien (13 January 1864 – 30 August 1928)

36

Next, making some arbitrary but plausible assumptions which enabled the Maxwell's law of distribution to be applied to the molecules of the black body, Wien obtained the following expression for the energy distribution:

$$dE = K\lambda^{-5} e^{-a/\lambda T} . d\lambda \quad \text{(Wien's formula)}$$

where K and *a* are constants.

The radiating capacity J of a black body at absolute temperature T (that is the energy it emits in the form of light and heat every second) is proportional to the fourth power of T, as described by **Stefan–Boltzmann law**.

$$j^* = \sigma T^4.$$

This law was the first effort to relate the molecular properties of matter to the energy of emitted or absorbed radiation. Missing was the spectral characteristics of such radiation. For, if one needs to design a thermal radiometer to measure the temperature of sun or of a hot furnace, he could only interrelate the quantity of energy flow to the temperature of the hot emitter, but could not define with any probability the wavelength distribution of such measured radiation. Therefore, Stefan–Boltzmann law was supplemented with another law that could describe the peak of radiance distribution with the wavelength of radiation. **Wien's displacement** was, then, the only available law that interrelated the increase of temperature of a black body to the maximum wavelength corresponding to maximum brightness of the light emitted.

Both maxima, of brightness and its corresponding wavelength shifted or displaced towards the violet region of the spectrum. It is written as follows.

$$\lambda_{max} = \frac{b}{T}$$

Physicists now had at their disposal two universal laws of thermal radiation that could be applied to all bodies without exception. The first gives a correct description of increasing brightness of luminescence as a body is heated. The Wien law only speaks of color corresponding to maximum brightness of light radiation. It is tacitly assumed that in addition to this radiation, there remain the radiations of longer wavelengths, i.e., of a different color, that had started earlier at a lower temperature. When a body is heated, its radiation widens the spectral range, opening up fresh regions of the spectrum. As a result, if the temperature gets high enough, we have a complete visible emission spectrum. The white light is the whole spectrum at once.

Apparently, Boltzmann-Wien refrained from expressing the radiation energy in terms of the wavelength of the emitted light since such connection would encroach on the wave theory of electromagnetism in such intimidating manner that would have caused unpleasant consequences. Nevertheless, those consequences were unavoidable as nature dealt a blow to the investigators of thermal radiation from quite a different angle.

Rayleigh-Jeans Law
In order to express Boltzmann's energy radiant of a heated body in terms of the wavelength of the emitted radiation, the English physicists Rayleigh and Jeans proposed a unified law stating that the intensity of radiation emitted by a hot body is directly proportional to the absolute temperature and inversely proportional to the square of the wavelength of the emitted light.

$$j(\lambda) = \frac{8\pi k T}{\lambda^2}$$

This law appeared to be in good agreement with experimental findings. But it was suddenly discovered that the agreement was good only for the long-wave portion of the visible spectrum, the green, yellow and red. The law broke down as the blue, violet and ultraviolet rays were approached.

From the Rayleigh-Jeans law it followed that the shorter the wavelength, the greater should be the intensity of thermal radiation. Experiment failed to confirm this. What is more, a very unpleasant thing was that as we move to shorter and shorter wavelengths the **radiation intensity was supposed to increase without bound**. Of course, this doesn't occur. There can never be an unbounded growth in wave intensity.

This curious situation that arose in the theory of radiation became known as the 'ultraviolet catastrophe'. That was at the end of nineteenth century. Classical Mechanics failed to account for the proper relation of the frequency of radiation from a heated body and intensity of radiation. It was Max Planck who spotted the problem of the energy continuum of Maxwell's wave theory that conflicted with the discrete light spectrum of elements. The resolution of the conflict of Rayleigh-Jeans law with experiment came when Max Planck, who in 1900 introduced the concept of quanta, and by Albert Einstein, who in 1905 advanced the theory of corpuscular photons.

The trend of growth of ideas advanced in such slow rate due to the constant check of the balances between **the formulation of the theory of probabilities within statistical mechanics, the gathering of experimental data in the extreme ranges of the frequency of radiation**, and **stringent compliance with the laws of conservation**. Each of those three bodies of generation of ideas and subsequent verification of their conformity is credited for giving birth to the new laws of Quantum Mechanics.

CHAPTER 2:
THE BIRTH OF THE QUANTA

Thermal Radiation At High Frequencies

As we have seen, there were two laws dealing with the thermal radiation of hot bodies. Separately, they held true very well, but when joined into a single law it confronted the 'ultraviolet catastrophe'.

$$\text{Stefan–Boltzmann law:} \qquad j^* = \sigma T^4$$

$$\text{The Wien's displacement law:} \qquad \lambda_{\max} = \frac{b}{T}$$

Planck was guided by the simple idea of Rayleigh and Jeans who combined the two laws of thermal radiation into one and had obtained an absurd result for short wavelengths. It was clearly evident that Rayleigh and Jeans' conception of a unified law was a valid approach since the wavelength of radiation was independent from the absolute temperature, T. Thus, there must be something else wrong other than the two above laws.

For the then available experimental data of the light spectrum of elements, Planck must tie the frequency of radiation to its intensity such that the theory agrees with experiment. That is, both the Stefan-Boltzmann law and the Wien law must hold true, in addition to accounting for the ultraviolet behavior of the intensity of radiation. The formula should not have any 'infinities'. Planck's formula differed from Rayleigh-Jeans as follows.

$$j(\lambda) = \frac{8\pi h}{\lambda^2} \frac{1}{e^{h\upsilon/kT} - 1}$$

Where h is Planck's constant, k Boltzmann's constant, and frequency of radiation is related to the wavelength and speed of light by $\upsilon = c / \lambda$.

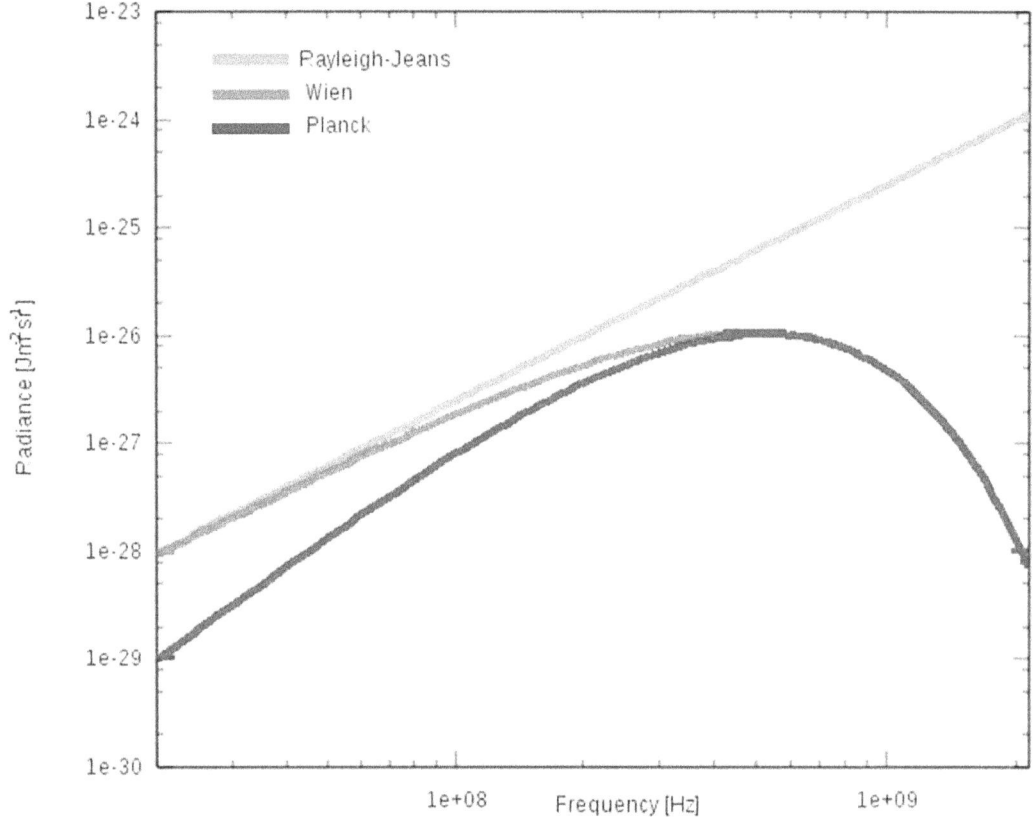

Comparison of Rayleigh–Jeans law with Wien approximation and Planck's law, for a body of 8 mK temperature.

Planck's formula, despite being a miraculous fit with experimental data, it defied the classical theory which presumes that energy is *continuous*. Planck recognized his victory since Classical Mechanics was clearly at fault in that exact spot. The spirit of Classical Mechanics recognized the discontinuity of things as an underlying principle but it failed to do so in defining discrete energy units. All objects have to be separated from one another and have boundaries. Objects do not pass one into another in continuous fashion, each one ends at some point. The molecules had clear-cut boundaries, and only the void between them was continuous. Incidentally, molecules somehow managed to interact through this emptiness.

Since the time of Faraday, Classical Mechanics had been trying to account for this interaction of bodies at a distance. **Aether** was the only available conceptual medium, via which the mutual action effects of the molecules. But, the industrial age had been erected on the ashes of the ancient fictional beliefs that shacked man to the horse-drawn wagons for countless number centuries. It was impossible to get the nineteenth century thinkers to retreat to the ancient ideas of unfounded conclusion such as the mysterious **aether**.

The Quanta

Molecular Energy Exchange:
It was held that when molecules collide, energy is exchanged in every imaginable quantity. This exchange followed exactly the laws of billiard balls. A moving molecule hits a stationary one, gives up part of its kinetic energy, and the two molecules then move off in different directions. In a head-on collision, the incident molecule can even come to a stop; then the struck molecule will fly off with the speed of the first one. Molecules are constantly exchanging energy. Molecules do not fuse or penetrate each others, they remain separable despite having variable energies. It was also clear that such **exchange of energies was discrete**, bounded, and could be described with greater mathematical precision in terms of conservation of momentum, mass, and energy.

Radiation Energy Flow:
Another form of energy exchange, that is not obviously connected with molecular motion, was the energy of wave motion. Since Maxwell proved that light is electromagnetic waves, the energy of light radiation must follow the laws obeyed by all waves. Again, this **energy is continuous**. It is propagated together with the moving wave flowing like water. Any given quantity of energy is consumed continuously in the same way that water continuously and indivisibly fills a vessel. James Clerk Maxwell derived it again using hydrodynamics in his 1861 paper "On Physical Lines of Force" and it is now one of the Maxwell equations, which form the basis of the wave theory of electromagnetic fields.

Now, there was no such notion of discreteness. It appeared that the ancient atomistic structure of matter did not demand that energy be composed of 'pieces'. It was enough to look around us to see that that was so. The light from a candle filled a room with an even flow of radiant energy, just as the sun kept up an uninterrupted stream of light.

The contrast of the molecular exchange of energy and the continuous flow of radiation energy tipped Planck towards the discrete molecular exchange. Planck concluded that **the energy of radiation (like matter itself) is atomistic and that it is released and acquired not continuously but in small portions, quanta,** as Planck called them, from the Latin 'quantum' meaning quantity.

Now we come to the magnitude of these separate portions of energy. **Planck made the extremely important discovery that these portions differ for different types of radiation. The shorter the wavelength of light, that is the higher its frequency (in other words, the 'more violet' it is), the larger the portion of energy.**

Mathematically, this is expressed by means of the well-known Planck's relation between the frequency and the energy of a quantum:

$$E = h\,\upsilon$$

Here, E is the energy of the quantum; υ is the frequency of the quantum; $\hbar$ is a proportionality factor which turned out to be the same for all types of energy that we know. It is known as *'Planck's constant'* or the 'quantum of action'. The value of this number is just as great to physics as its magnitude is small:

$$h. = 6.62517 \cdot 10^{-34}\ \text{J} \cdot \text{s}$$

It is this insignificant magnitude of the quantum that makes the light of a candle or the sun appears to us to burn with a constant glow. In the classical sense, small and insignificant numbers were irrelevant to researchers. But even before Thomson's discovery in 1897 of the electron with it extremely small mass of $9.10938215(45)\times10^{-31}$ kg, Avogadro had introduced the constant $6.02214179(30)\times10^{23}$ mol^{-1} that describes the colossal number of molecules in the molecular weight of gas. Therefore, Planck's constant falls comfortably within the realm of small numbers that describe subatomic particles. That was followed latter by Millikan's accurate determination of the charge of the electron - $1.602176487(40)\times10^{-19}$ C. The smallness of numbers in the subatomic world continued with the Einstein's famous law: $E = mc^2$, that described the intangible amount of mass gained or lost during the absorption or emission of radiation.

To illustrate, let us calculate the number of quanta radiated by a 25-watt electric light bulb per second. Taking the emitted light to be yellow (wavelength 600 nm) , we find by Planck's relationship that:

Number of quanta emitted from 25-watt lamp = 25 $(c/\lambda)/\hbar$

$N = 25\ watt\ /\ [[(3x10^8\ m/s)\ /(600\ x\ 10^{-9}\ m)]\ (6.62517\bullet.10^{-34}\ J\bullet s)] = $ 7.57903E+19 quanta of yellow light per 25 watt. which is 60 million x million x million portions of energy per second. All of that is radiated by a small 25-watt bulb every second. Quite obviously, the human eye is not sensitive to slight fluctuation in such magnitudes of energy.

On December 17, 1900, Planck addressed the impasse of Maxwell's theory of electromagnetism in regard to the *discreteness of energy*. He attempted to overcome one of the difficulties of the theory of thermal radiation such that the spectra of the emitted lights would fit the discrete ranges of frequencies encountered in experiments.

Planck noted that Boltzmann's Equation was written in such precise manner that defied finding flaws in it. Thus, Planck's only option was to rewrite the same equation in a manner amenable to experimental interpretation. Planck relied on pure theoretical analysis by arranging the terms in Boltzmann equation in such as manner that yielded a universal constant of the discrete amount of energy that could be transported due to transition from initial state to final state as follows.

$$E(\text{photon}) = h\nu = \frac{hc}{\lambda} = E_f - E_i$$

Planck's constant, or the quantum, was discovered or invented on that date of December 17, 1900. However, Planck never contemplated that the *quanta* would be attached to photos in the context of *particles*. That would be introduced latter by Einstein. Planck's constant became a universal constant in physical sciences.

Max Planck (April 23, 1858 – October 4, 1947), Germany.

Photons And Corpuscles
Planck started the quantum age without knowing that Einstein would soon introduce the boldest idea yet, that **light emits from burned mass**. Planck did not initially tolerate Einstein's eccentric twist of the quanta that suggested that a tiny amount of mass must be converted into quanta in order to produce light. Even though Einstein's daring idea of conversion of mass into light was a direct result of Newton's law that related energy to the square of the speed and the mass contents of the object, yet Einstein dared to link Planck's quanta to an unforeseen amount of burned mass and to the number of emitted photons.

Initially, Einstein conjectured that the energy, E, of the light emitted from a heated object must be equal to mV^2, where V was latter changed to the conventional c, the speed of light and m was the newly introduced mass loss, which Einstein proposed. Einstein argued that the extreme smallness of the consumed mass m explains its inaccessibility to detection. Of course, the classical "½" that precedes mV^2 was not needed for photons moving at the speed of light, from the start to end. In other words, once mass converts to lights, photons rush at the full speed of light without having to gradually accelerate, as inertial matter does. The law is now written as $\Delta E = c^2 . \Delta m$.

Hence, a new bridge is erected between the unforeseen subatomic interior of matter and the elusive photons, despite the fact that both sides of the bridge were inaccessible to direct measurements. It should however be stated that both Planck's and Einstein's new theories have grown out of Newton's Classical Mechanics. That was later criticized by Dirac as an inherent limitation of Quantum Mechanics that grew out of its old predecessor.

Einstein's new law of conservation of subatomic mass into radiation energy became the governing canon in the new fields of radioactive isotopes, semiconductors, elementary particles, masers, lasers, and nuclear reactors. The oldest of those new fields is hardly seventy years of age. They are all off springs of twentieth-century physics. In this age, knowledge is advancing at an exponential rate. Every new discovery or invention opens up newer frontiers. Physics was thrust into the unknown with the invention of high energy accelerators, outer space research, and multi-channel energy analyzers and spectrometers.

The new fictional mechanics works in four dimensional spaces where mass and time are no longer absolute constants and where discreteness characterizes every intervening variable. Nothing is limitless or boundless except the space of the universe. The emptiness of space and its ethereal intrigue will be redefined in terms of matter and antimatter balance and inhomogeneity when gravitational forces bend the geodesics of space.

Now, as the focus narrows on the nature of photons, their entity became the subject of more precise scrutiny: Whether photons were indivisible tiniest particles or, solid impenetrable spheres? How many different varieties of such tiniest spheres are there? The birth of photons is now explained by Einstein, their individual identity by Planck. Further, the speed of photons was proven to be remarkably constant regardless of the motion of the emitting source. It will be another three decades before Dirac redefines the emptiness of space in terms of negative rest-energy.

In Classical Mechanics, objects do not change their state of motion unless encountered force. But light could emit from objects at the exact constant speed regardless from the speed of the object. *Such stark violation of the classical laws of mechanics is still unresolved even by Quantum Mechanics.*

In addition to its stubbornly constant speed, light emits from heated bodies at predicable frequencies and intensities that fitted no classical canon. This had the most shattering effect on Classical Mechanics. During the last years of its undivided rule, the mysterious processes of radioactivity, not only smashed atomic nuclei, but exploded the very basis of physics, those principles that had appeared so obvious from the standpoint of experimental detection. Out of these cracks in the structure of Classical Mechanics grew the Theory Of Relativity and the Quantum Theory.

The new era of Quantum Mechanics opened a new world where sensual observations must be harvested through ever complicated and ever growing numbers of new instruments. The naked eye watch was no longer capable of tracking the fast speeds, the extremely short times and sizes of the new objects.

Lorentz Transformation

Initially, Newton's classical mechanics described the relations between coordinates and time of an event viewed in two frames S and S', where S' moves at constant speed v along x-axis of frame S, by **Classical Transformation**:

$x' = x - v t$ ....(1-a)
$y' = y$ ....(1-b)
$z' = z$ ....(1-c)
$t' = t$ ....(1-d)

Between 1881 and 1887, Michelson and Morley attempted to prove the validity of the above classical transformation by direction of earth's rotation, east-west as frame S' moving as speed v from the stationary south-north frame; S. They obtained surprisingly unexpected results; at different times of the year, when the directions of the earth's orbital velocity are different, always gave the same negative result. These experiments were performed at the Norman Bridge Laboratory, Pasadena. Therefore, the expected effects of the motion of the earth through an ether at rest could not be found, motion through the ether could not be detected by optical experiments.

In 1887, Lorentz, in order to explain the negative results obtained by Michelson and Morley and to allow for the constancy of the speed of light c in the

Hendrik Antoon Lorentz(18 July 1853 -4 February 1928), Netherlands.

two frames (S as the South-North and S' as East-West), modified the above equations to the following *Lorentz Transformation*:

$x' = \gamma (x - vt)$ ....(2-a)
$y' = y$ ....(2-b)
$z' = z$ ....(2-c)
$t' = \gamma (t - xv/c^2)$ ......(2-d)

where γ is the famous *Lorentz factor*, which can be derived by simultaneity of events as shown below. Obviously, Lorentz obtained equation (2-d) by dividing (2-a) by c and substituting by $t = x/c$. This is very important assumption as it relies entirely on the *relative simultaneity* of events defined only in terms of the optical signal traveling at speed c in both S and S'. In frame S, the event that occurs at it origin O is observed at time $t = x/c$, while in frame S', the same event at O is observed at time $t' = \gamma (t - xv/c^2)$.

To prove that $\gamma = 1/(1 - v^2/c^2)^{1/2}$, Lorentz relied on his mathematical skills by assuming that the event in the first frame is observed according to the equation:

$x - ct = 0$ ....(3-a)

while in the moving frame, the same event is observed by the equation:

$x' - ct' = 0$ ....(3-b)

In order to account for symmetry of space, Lorentz must prove that the direction of motion of the frames should NOT matter. Thus, Lorentz tied the above two equations as follows:

x' - ct' = λ (x - ct) (positive direction of x-axis) ....(4-a)
x' + ct' = μ (x + ct) (negative direction of x-axis) ....(4-b)

where λ and μ are constants of symmetry, to be determined.
Solving the above two equations for x' and t' and using v = x/t, Lorentz obtained the formula (omitting many steps in the proof)

$$\gamma = 1/(1 - v^2/c^2)^{1/2} \ ....(5)$$

which accounts of the two constants λ and μ.
$$\gamma = (\lambda + \mu) / 2$$
$$ß = (\lambda - \mu) / 2$$
with ß/γ = x/ t c = v / c

$$ß = \gamma \ v/c = (v/c)/(1 - v^2/c^2)^{1/2} \ ....(6)$$

Now, in 1887, Lorentz had mathematically tamed the constancy of the speed of light, in two frames of reference moving relative to each other at constant speed, through his the factor $\gamma = 1/(1 - v^2/c^2)^{1/2}$. A new era of relativistic views of the universe was brewing from 1887 until 1905. Scientists must account for the eccentric and strange nature of light and the mathematical gimmicks of Lorentz. Both have breached Classical Mechanics. Both violated the properties of rigid bodies or rigid time scale.

Discovery Of Photons

In 1905, Albert Einstein postulated that light was made out of corpuscles carrying equal amounts of energy corresponding to Planck's quanta and which were responsible for dislodging electrons from the surface of illuminated metal. Einstein published his theory of the photoelectric effect in metals in the German journal "Physikalische Rundschaa". At the time that Einstein took up this study, the effect was well on in years. Naturally, there was no current through the airless space around a metal plate. But when the light of a mercury lamp was made to fall on one of the plates, current immediately began to flow in the electric circuit. When the light was turned off, the current stopped. It was clear that current carriers (electrons) originated only when the plate was illuminated.

It was quite obvious that these electrons were ejected from the illuminated metal much like molecules jump into the air from the surface of heated liquid. But, light is an electromagnetic wave. It is difficult to imagine how a wave can knock electrons out of metal. There is no collision here of energetic molecules, as a result of which one of them is ejected from the surface of a liquid.

Albert Einstein (14 March 1879 – 18 April 1955), Germany.

Another interesting circumstance was noted. For each metal studied, there appeared to be a certain limiting wavelength of incident light. When this wavelength was exceeded, the electrons in the flask disappeared at once and the current ceased to flow no matter how strong the light was. This was altogether strange.

It was clear that electrons are ejected from the metal because the light in some way conveys energy to them. The brighter the illumination is, the stronger the current. The metal receives more energy and larger quantities of electrons can be knocked out. But no matter what the wavelength of the light, the metal should be receiving energy all the same. True, with increasing wavelength the energy diminishes and fewer electrons are ejected from the metal, but still there should be some kind of current.

Yet experiment showed no current at all. One would think the electrons ceased to accept the radiant energy. How was one to figure out why electrons were so particular about the energy they were given? That was something that the physicists just could not grasp.

Einstein regarded the photoelectric effect from a different angle. He attempted to picture the actual process of the ejection of an electron from a metal by light. In normal conditions, there is no cloud of electrons hovering over the metal. Electrons are bound to the metal by some kind of force. To knock them out of the metal, a little energy is needed.

In Stoletov's experiments this energy was supplied by light waves. But a light wave has a definite wavelength, something on the order of a fraction of a micron, and its energy is, as it were, concentrated in the minute volume occupied by an electron. This means that in the photoeffect a light wave behaves like a tiny '*particle*'. It strikes an electron and dislodges it from the metal. Then what would the energy be of such a particle? Calculations show that it would be very small. Then why not suppose that it would be exactly equal to the quantum that Planck had conjured up five years before?

Einstein postulated that light is simply a stream of quanta of energy, all the quanta of a single wavelength being exactly the same, which is to say that the quanta carry identical portions of energy. Later, these quanta of light energy were given the name photon.

A photon carrying a small portion of energy strikes an electron with sufficient force to knock it out of the metal. On the other hand, obviously, if the photon energy is insufficient to disrupt the electron bonds in the metal, the electrons will not be knocked out and there will be no current.

According to Planck's formula, the energy of a quantum is determined by its frequency, and the greater the wavelength of the light, the lower the frequency. Hence it is quite obvious that the photoelectric effect has definite limits. It is simply this: **If the wavelength of the light was too large, the photons do not have energy enough to dislodge electrons from the metal.**

What is more, it doesn't make any difference how strong the light is, whether a thousand or only two photons strike the metal and bombard its electrons: the latter are indifferent. The situation changes if the photons have sufficient energy. In this case, the brighter the light, the more photons enter the metal every second, and the greater the number of electrons ejected, thus producing a stronger current.

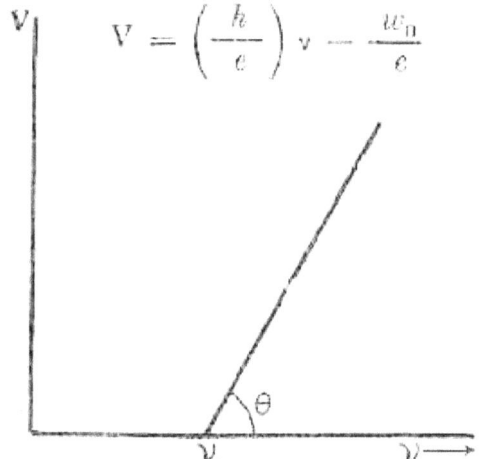

$$V = \left(\frac{h}{e} \right) \nu - \frac{w_0}{e}$$

Relation of cut-off voltage with the frequency of light in experiment of photoelectric effect. V is the voltage that suffices stopping electrons from jumping of the metal surface when irradiated by light of varying frequencies. No matter how intense the light is, V is zero below the threshold frequency of radiation.

Dual Nature Of Light

Both the wave and the corpuscular nature of lights have appeared in physics at about the same time. Newton suggested that bodies shine by ejecting streams of light particles or corpuscles. Huygens opined that bodies shine by pulsating and forming waves in the surrounding aether. Finally, at the beginning of the nineteenth century the experiments of Young, Fresnel and Fraunhofer resulted in what would have seemed a decisive victory for the wave theory of light.

The newly discovered phenomena of *interference, diffraction* and *polarization* of light were in excellent accord with Huygens' theory and quite incomprehensible from Newton's viewpoint. Optics began to develop. Brilliant optical theories were developed and complex optical instruments were constructed. Finally, Maxwell completed the structure of optics by proving the electromagnetic nature of light waves. The triumph of the wave theory was complete and indisputable. But less than fifty years passed and the corpuscular theory of light was again revived. The photoelectric effect which the wave theory had failed to explain-what an annoying blemish on an otherwise perfect structure-was accounted for in amazing fashion by the opposing theory.

The century-old argument again flamed up. Gradually it dawned on physicists that the amazing and inevitable view had to be that light is at the same time both waves and particles. But **why is it that light never manifests itself completely in this twofold manner?** Sometimes it appears only as particles, yet at other times it is only in the form of waves.

The second question that came with Einstein's theory was not simple either. It appeared that in the photoelectric effect **the electrons did not react to just any portion of energy offered them. The portion of energy had to be of a very definite magnitude or greater, otherwise the light energy found no response**.

Joseph von Fraunhofer (6 March 1787 – 7 June 1826), Germany.

It also turned out that an electron which is not bound by any forces to neighboring ones ceases to be particular and responds to all kinds of energy packets.

Rutherford Atom Model

Thomson's plum pudding model of the atom has already collapsed after Rutherford detected the heavy center of the atom. In 1911, Ernest Rutherford proposed a new model of the atom after observing the scattering trajectories of alpha particles by various atoms. He bombarded atoms of various substances with the newly discovered alpha rays of radioactive substances. It was already known that these rays consist of positively charged particles. Studying the scattering of alpha particles by atoms, Rutherford concluded the alpha particles were scattered as if they were repulsed not by the entire positive cloud of the Thomson atom, but by a very small portion of the atom concentrated somewhere at the center. Rutherford found in the nucleus of the atom the analogy with the planets and stars. In the new model of the atom, the minute particles-electrons- are revolving in definite orbits around the heavy nucleus. The dimensions of these orbits are tens of thousands of times greater than those of the electrons and nuclei. Thus, the atom was an *'empty'* structure, as was the planetary system where the dimensions of the planets are thousands of times smaller than those of the orbits about the sun.

The entire positive charge of the atom appeared to be concentrated in this tiny central part. Rutherford called this part of the atom the core (nucleus). Then where are the electrons? The old view that the electrons were bound to the positive charge in the atom by the electric forces of attraction was not in doubt. But since the electrons exist at a certain distance from the core, there must be some force that counterbalances the electric force of mutual attraction of electrons and nucleus. It was obvious that this force had to be operative all the time.

He became known as the father of nuclear physics. In 1911, Rutherford postulated that atoms have their positive charge concentrated in a very small nucleus and thereby pioneered the Rutherford model or planetary, model of the atom through his discovery and interpretation of the scattering alpha particles in his gold foil experiment. He is widely credited with first splitting the atom in 1917, and leading the first experiment to split the nucleus in a controlled manner by two students under his direction, John Cockcroft and Ernest Walton in 1932.

As it stood, Rutherford's atom model reconciled the available experimental facts that electrons were easily removed from the atom than the nucleus, and were the outer bond-forming atomic particles, and nucleus was essentially extremely small and central in the atom. Atoms exist for a sufficiently long time, and so the countering force would obviously have to be just as constant as the force of electrical attraction between the electrons and the nucleus. It seemed reasonable to think that this was a centrifugal force. It appears as if electrons revolve about the atomic core. It could be calculated whether the force is sufficient to keep the electrons from falling into the nucleus. Calculations showed that it is quite sufficient if the electrons revolving about the nucleus move at speeds of many tens of thousands of kilometers per second and at a distance from the nucleus of the order of hundred millionths of a centimeter. This was the Rutherford model of the planetary structure of the atom.

Ernest Rutherford (August 30, 1871- October 1937), New Zealand.

The motion of electrons about the nucleus is accelerated motion (the electrons move along closed curves). Hence, *there must be electromagnetic radiation.* The classical laws are equally applicable to the Thomson model and the Rutherford model of the atom. But, unfortunately, the success is also the same. In radiating light, the electron uses up its energy. In doing so, it slows down in millionths of a second and must inevitably fall onto the nucleus, just like a satellite decelerated in the earth's atmosphere falls to earth. The fate of the electron should be the same as that of the satellite. An atom, under such conditions, would very soon cease to exist. But atoms live on. Electrons should not be giving up energy and should not emit light. But bodies do emit light when heated.

Rutherford's atom model shows the enormous magnitude of Coulomb's force that binds electrons to the nucleus as follows:

Electron Charge (e) = $-1.60219*10^{-19}$ coulombs
Electron mass (m_e) = $9.10953*10^{-31}$ kgs
Bohr Radius (r_o) = $5.29177*10^{-11}$ meter
Earth Gravity (g) = 9.80665 meter/ sec ^{2}
Coulomb's Force on inner electrons (g) = (e^-e^+ / r_o^2)/ ($m_e g$)

= $[(-1.60219*10^{-19})(1.60219*10^{-19}) / (5.29177*10^{-11})^2]/ (9.10953*10^{-31} * 9.80665)$

= $-1.02615* 10^{+12}$

Thus, the Coulomb's force is equivalent to more than *million million times the gravitational force* on the surface of the earth.

Bohr Atomic Model

Niels Bohr tackled the two problems that confronted the Rutherford planetary model of the atom while attempting to understand the origin of the discrete spectra of elements in relation to heating their atoms. Both Rutherford's and

Thomson's models of the atom suffered from the imbalance of energy due to the curved motion of electrons in the atom. Neither model was able to corroborate the mechanism of absorbing or emitting radiation in the specified and discrete fashion observed in the spectrum of the heated elements.

By the years 1911, Classical Mechanics was not able to account for the luminescence of heated bodies, and it could not explain the existence of discrete and element-specific spectra. Bohr must find more than just the electron deceleration mechanism in order to account for the specificity of the spectral wavelength to specific chemical elements. Bohr found himself facing the exact challenge of energy continuum that was rejected by Planck in 1900 and Einstein in 1905. The two models of the atom perceived electron in dynamic gyration around the nucleus but could not explain why electrons do not luminesce when not heated, do not fall into the nucleus and disappear outright. And when heated, atoms emit radiation in quantized fashion characteristic of their structure.

He proposed discrete states where electrons move around the nucleus without radiation and that only the change of those states would cause the absorption or emission of radiation.

Further troubles with Classical Mechanics came from the special theory of relativity. Einstein has already attacked two classical taboos: the constancy of time and mass. That set the stage for others to reexamine all classical postulates with critical eyes.

Bohr rejected the concept that an electron in an atom need not give off light even when in accelerated motion and postulated the existence of specific orbits about the nucleus where electrons move in steady state energy levels.

He then advanced a second postulate. **Let us suppose an electron in orbit suddenly jumps to another orbit of less energy. Where has the excess energy gone? Energy cannot simply vanish away into nothing. Seek it outside the atom, says Bohr.**

This energy is ejected from the atom in the form of a quantum, that same quantum of light energy which Einstein called a photon. An electron that has emitted a photon takes up a different orbit and does not emit light any more. The photon was ejected during the minute fraction of time when it jumped from one orbit to the other. Meanwhile the photon was making its way through the other atoms and finally got out of the substance.

Niels Henrik David Bohr (October 7, 1885 - November 18, 1962), Denmark.

According to Planck, the energy content of a quantum depends on the wavelength of radiation through the proportionality constant, h. Hence, Bohr's emitted quanta should decipher the magnitude of the energy level of Bohr's postulated orbits and the wavelength of the spectral line recorded on the photographic plate during the heating of the emitting element. The photons emitted during jumps between two orbits are all the same.

$$E_f - E_i = \Delta E\,(\upsilon) = h\,\upsilon$$

The blackness on the plate at the site of this spectral line indicates the number of photons impinged on the plate; the more there are, the blacker the line. The more photons, the brighter the body that has emitted them. All the atoms of a certain substance are exactly alike. Hence, the electrons all exist under the same conditions. All the transitions that electrons make between two orbits yield, in the final analysis, a single unique spectral line.

An electron can reside in any one of them, in turn. Every jump from a higher-energy orbit to one of lower energy is accompanied by the birth of a photon. But since there is a difference of energy between different orbits, the photons will have different energy and frequency. A photographic plate will then exhibit a series of narrow spectral lines. This is exactly what the spectrum of gaseous Hydrogen looks like. It has several tens of lines with different wavelengths.

Generally speaking, such a simple spectrum as that of Sodium consisting of only one line is a rarity. Spectra usually have many tens of lines and frequently even thousands of lines. The spectral patterns of some chemical compounds are so intricate that there doesn't seem to be any hope of disentangling them. But there are laws to go by which make the task easy.

By examining a spectrum, one can draw all manner of conclusions about the conditions under which atomic electrons exist. Just about all that we know about the electron shells of atoms has been acquired through a painstaking analysis of their spectra.

Privileged Atomic Quantum Orbits
Bohr's steady-state atomic shells remedied the flaw of the classical approach to subatomic motion as follows.
Consider the general case of a linear simple harmonic oscillator with displacement x, at time t, given by

$$x = A \sin (2\pi v\, t)$$

where A is the amplitude and v the frequency. As the total energy of the oscillator changes from all kinetic, at the equilibrium position to all potential, at the maximum displacement, it can be determined by computing the kinetic energy at the equilibrium position. The kinetic energy of the oscillator at the instant t is

$$\text{K.E.} = \tfrac{1}{2}\, m\, v^2 = \tfrac{1}{2}\, m \left(\frac{dx}{dt}\right)^2$$

where m is the mass of the oscillator and $v = dx/dt$ its linear velocity at the instant considered. At the equilibrium position dx/dt is maximum

$$W = -\frac{1}{2}\, mv^2 N.$$

$$= \frac{1}{2}\, m\, (2\pi v A)^2 = 2\pi^2 v^2 A^2 m$$

According to the quantum theory, this energy should be **an integral multiple of hv.**

$$\therefore \qquad nhv = 2\pi^2 v^2 A^2 m, \quad \text{where } n \text{ is an integer,}$$
$$\text{or} \qquad nh = 2\pi^2 A^2 vm$$

The momentum p_x of the oscillator at the instant t is given by

$$p_x = m\,(dx/dt) = m.2v\pi A \cos 2\pi vt$$
$$\text{Putting} \quad m\,.\,2\pi v\, A = B, \quad p_x = B \cos 2\pi vt$$
$$\therefore \qquad p_x / B = \cos 2\pi vt$$
$$\text{From eqn. (1)} \quad x/A = \sin 2\pi vt$$
$$\therefore \quad x^2/A^2 + p_x^2/B^2 = 1$$

Thus, the relation between p_x and x is given by an **ellipse**. If we draw a graph with x as abscissa and p_x as ordinates, it will be an ellipse whose semi-major and semi-minor axes are A and B respectively.

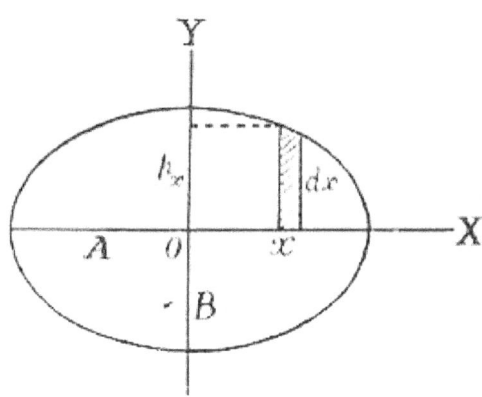

Considering such an ellipse, let dx be the width of an element at a distance x from the origin. Let p_x be the value of the ordinate corresponding to x. Then the area of the element considered is $p_x \cdot dx$. The area of the ellipse is obtained by the integration of $p_x \, dx$ over a complete cycle, which is known as the "phase integral", represented

$$\therefore \qquad \oint p_x \cdot dx = \text{area of the ellipse}$$
$$= \pi \times A \times B$$
$$= 2\pi^2 A^2 \nu m$$
$$= nh$$

Thus the phase integral of a linear oscillator is an integral multiple of h. the Planck's constant. Applying this result to the electron in uniform circular motion, which is equivalent to a harmonic oscillator, and replacing the linear momentum p_x. by the angular momentum p_φ and the linear element dx by the angular element $d\varphi$ appropriate to the case, **we** can write

$$\oint p_\varphi \cdot d\varphi = nh$$

The angular velocity being assumed constant, p_φ is also constant.

$$\therefore \qquad \oint p_\varphi \cdot d\varphi = v_\varphi \cdot \oint d\varphi = p_\varphi \int_0^{2\pi} d\varphi = p_\varphi \cdot 2\pi = nh$$

$$i.e. \qquad p_\varphi = n\,(h/2\pi)$$

Hence; according to the quantum theory, the angular momentum of the electron moving in a circular orbit can have only those values which are integral multiples of $h/2\pi$, or conversely **the permitted circular orbits of the electron are those in which the angular momenta are integral multiples of** $h/2\pi$

In order to derive expressions for the radius, frequency and energy of the permitted orbits of the electron in the hydrogen atom, let M and E be the mass and charge of the nucleus; m, e and v the mass, charge and linear velocity respectively of the electron which is assumed to revolve in a circular orbit of radius a.

In general, the nuclear charge $E = Ze$, where Z is the atomic number of the element; for hydrogen: $Z = 1$ and $E=e$.
The nuclear mass M is so large compared to the electronic mass m that, for the present, the nucleus is assumed to remain at rest.
The electrostatic force of attraction between the nucleus and the electron $=Ee/a^2$.
The centrifugal force of repulsion between the two resulting from the circular motion of the electron = mv^2/a.
The system will be stable, if

$$Ee/a^2 = mv^2/a$$

Introducing the quantum condition for the orbit,

$$p_{\varphi} = I\omega = n\,(h/2\pi)$$

But
$$I\omega = ma\omega^2 = ma^2 \cdot v/a = mav$$

$\therefore$
$$mav = nh/2\pi$$
$$v = nh/2\pi ma$$

Which gives the radius of permitted obits as

$$a = \frac{nh}{2\pi m} \times \frac{nh}{2\pi Ee} = \frac{n^2 h^2}{4\pi^2 Eem}$$

Thus, the **radius** a of the permitted orbit is directly proportional to n^2, since all the other quantities are constant. This means that the radii of successive permitted orbits are proportional to the squares of the integers 1, 2, 3,... These integers are called the **quantum numbers** of the respective orbits.

The radius of the first smallest orbit in the hydrogen atom can be readily calculated from the above relation by using the known values of h, E=e, and m and putting it n =1. It turns out to be equal to 0.53×10^{-8} cm, and is known as the **Bohr radius**. The diameter of the first orbit is, therefore, of the order of 10^{-8} em., which agrees with the values of the diameters of atoms computed by various other methods.

The orbital **frequency**, f, is given by

$$f = \frac{\omega}{2} \cdot \frac{v}{2\pi a} = \frac{2\pi Ee}{nh} \cdot \frac{1}{2\pi a} = \frac{Ee}{nha}$$
$$= \frac{Ee}{nh} \cdot \frac{4\pi^2 Eem}{n^2 h^2} = \frac{4\pi^2 E^2 e^2 m}{n^3 h^3}$$

According to the classical theory, this orbital frequency is equal to the frequency of the spectral line emitted by the atom. But we shall presently see that it is not so according to Bohr's theory.

The total energy W of the electronic system is equal to the sum of the kinetic energy ($0.5\ mv^2 = Ee/2a$) and potential energy ($-Ee/a$).

$\therefore$
$$W = \frac{Ee}{2a} - \frac{Ee}{a} = -\frac{Ee}{2a}$$

or

$$W = W_n = -\frac{Ee}{2} \cdot \frac{4\pi^2 Eem}{n^2 h^2} = -\frac{2\pi^2 m E^2 e^2}{n^2 h^2}$$

W_n being the energy of the electron when it is in the *nth* orbit.

In this relation, since all the quantities except *n* are constants, the orbital energy is inversely proportional to the square of the quantum number of the orbit. Evidently *for any one particular orbit the energy is constant,* which means that **as long as the electron remains in that orbit it cannot lose energy by radiation**, in contradiction to the classical electromagnetic theory.

The interpretation of the negative sign associated with the expression for the orbital energy is important. As *'n'* increases, the absolute numerical value of the energy decreases, but **on account of the negative sign, the actual energy will increase**. This means that *the outer orbits have greater energy than the inner ones.* In the case of the hydrogen atom, taking the relation

$$W_n = -2m \left(\frac{\pi e^2}{h} \right)^2 \cdot \frac{1}{n^2}$$

Since the first orbit has the least energy it is the most stable and is the one which the electron occupies in the normal unexcited atom. Expressing the orbital energy in terms of the more convenient unit of volts is obtained by the relation: energy in ergs = eV/300, where V is in volts and e in e.s.u. (4.77×10^{-10})

$$V_n = \frac{-2m(\pi e^2)^2}{n^2 h^2} \times \frac{300}{e} = -\frac{13 \cdot 6}{n^2} \text{ volts,}$$

Emission And Absorption Of Radiation

In order to set reference for the direction of energy flow, Bohr postulated that the farther an electron is from the nucleus, the higher its total energy; the closer it is to the nucleus, the lower its energy. When an electron jumps to an orbit closer to the nucleus, it diminishes its energy, so that photons are emitted during such transitions. And on the contrary, the farther the orbit is from the nucleus, the closer the electron is to 'escaping' from the atom and the more energy the electron has. The sign of the magnitude of energy has been reckoned such that zero total energy corresponds to a free but stationary electron. Thus, a bound electron possesses negative total energy.

A free electron was viewed as state where an electron does not fall under the balance of Coulomb's electric forces between other electrons and the nucleus. A bound electron, on the other hand, must repulse other orbital electrons and attract the nucleus of the atom while in constant circular orbiting. Since all electrons in the atoms are in constant motion, the balance of Coulomb's forces is in a steady state of uniform motion that will only be described by rigorous differential equations.

Evidently, the electronic obits that are closest to the nucleus have minimal energy level and could not emit radiation and hence would not fall into the nucleus. Those are the most stable orbits, which contradicts the common sense of Classical Mechanics. In fact, even Bohr could not explain why the inner electrons that are subjected to the greatest Coulomb's attraction to the nucleus would not fall into it. And the only explanation given by Quantum Mechanics was in the form of a probability function that makes such electron absorption by the nucleus a rare event. Similarly, the probability function is a product of a mathematical model that was based on the same assumptions and data available to Bohr and Rutherford. Therefore, one should not expect that the probability function to predict unprecedented facts far and beyond what the wave equation was founded on.

When heated, the inner electrons acquire energy to jump upwards the ladder to the outer orbits. The difference in the jump from inner energy levels to outer energy levels depends on the amount of heat imparted on the electrons. In order to restore balance (electrical, dynamic, and energy balances), the electrons must find the atomic shell that fits their energy and spin conservation. Further, the electrons also must perform under the Coulomb's potential. The most optimum balance is therefore the release of extra energy acquired from heating and returning back into an inner orbit. **As the electron jumps inwards, a photon is ejected;** the heated body glows, or emits light. The body is now emitting light. The decelerated electron complied with Coulomb's attraction, restored its orbital spin conservation, and maintained the overall atomic balance of forces.

If greater amounts of energy are imparted on the electrons, the thermal motion of the atoms becomes more energetic, collisions are more frequent and violent. The electron spends only a little time in its innermost orbit. The atoms more and more frequently remain into an excited state then return to 'normal' only to leave it again almost immediately. At this point, photons are being generated by thousands and millions every second. They build up avalanche-like as the temperature rises which confirms the Stefan-Boltzmann law. But it is not only the number of photons that is increasing. The lengths of the electron jumps also increase. Jumping back from outer orbits, the electrons generate very strong photons at greater frequencies and the smaller wavelengths. The emitted light becomes brighter and more 'violet' in accord with Wien's displacement law.

Bohr's theory was thus able to account for the most subtle processes of emission and absorption of light by atoms and the detailed spectral structure of atoms and molecules.
But, where did the photons come from?

And how could an atom emit electromagnetic waves of very long wavelength, that far exceeds the dimensions of atomic orbits?

Further, the negative potential energy of bound electrons was a deliberate manipulation of the boundary conditions of of the harmonic oscillator that should raise some skepticism.

Brightness And Spin Of Emitted Radiation

From Bohr's theory we could find the wavelengths of the photons generated in electron jumps from orbit to orbit. But the theory was helpless as far as accounting for the brightness of the spectral lines was concerned. It was not clear how one could calculate the number of photons in the spectrum.

According to Quantum Mechanics, an electron in an atom radiates separate lines or, a discrete spectrum. The rungs of the energy ladder of electron orbits have different heights. The height is less, the farther the orbit is from the nucleus. Thus, long wavelength lines of the spectrum, which correspond to electron jumps between orbits distant from the nucleus, must be close to one another, which makes it appear as an almost continuous spectrum. Thus, the long wavelength section of the 'quantum' spectrum should not differ materially from the very same section of the 'classical' spectrum. In this region, where n is great, the brightness of the first spectrum could be calculated on the basis of Classical Mechanics. And then we could extend the calculation to the entire 'quantum' spectrum.

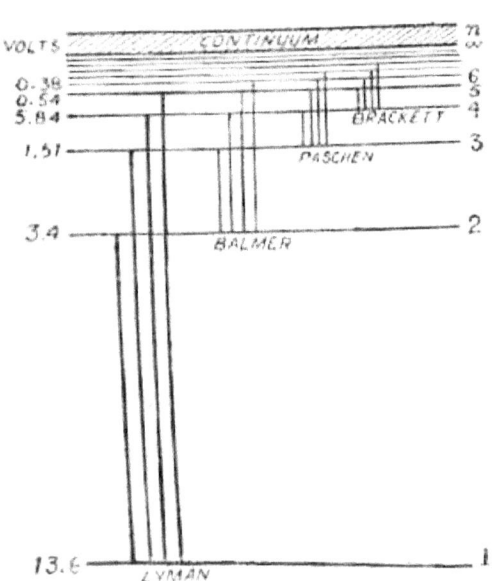

Privileged electronic levels of the Hydrogen atom.

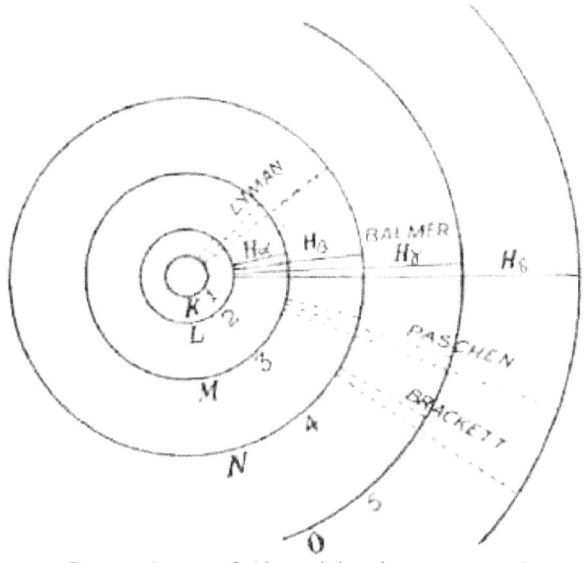

Spectra of the Hydrogen atom

That is the **correspondence principle**, which Bohr improvised to save his theory. It borrowed the continuous spectrum of energy from Classical Mechanics and attempted to extend it to Quantum Mechanics. It will soon crumble with the discovery of the spin. Experiment yielded line brightnesses that differed from those of Bohr's theory.

A closer look at the new theory will show that the correspondence principle was not the only lapse in Bohr's theory. Actually, from the very start all its basic premises bore clear traces of Classical Mechanics. Bohr rejected the classical views on electron motion, yet introduced the concept of electron orbits in the atom. He was firmly convinced that electrons revolved about the nucleus of the atom in the same way that the earth moves round the sun. Bohr 'prohibited' the electron from radiating while in orbit, but he could not find any good justification for doing so.

Bohr's theory gave a correct qualitative explanation of the origin of photons in atoms, but the process as such remained a mystery. It did not follow from any of its postulates. This dual nature of Bohr's theory was quick to manifest itself.

New facts cropped up that did not fit into the framework of the theory. Yet it had merits. Bohr's theory was a tremendous step forward in understanding the world of the atom. And it had its limitations. It explained much that was incomprehensible and beyond the means of Classical Mechanics. But almost, much remained unaccounted for. The time for new steps had come. And they were soon made. The first was taken by the French physicist Louis de Broglie.

de Broglie Matter Waves

de Broglie suggested that electrons, nuclei and generally all material building blocks' of our world possess properties of waves and corpuscles (particles). As a result, particles of matter, including atoms like those of light earlier, could no longer be visualized.

In 1924, Louis de Broglie postulated the possible existence of matter waves. de Broglie maintained that these waves are generated in the motion of any body, whether a planet, a stone, a particle of dust or an electron. Like electromagnetic waves, these waves are capable of propagation in an absolute void. Hence, they are not mechanical waves. But they are produced in the motion of all bodies, including those not charged electrically. Hence, they are not electromagnetic waves.

Since the ether was put to rest without a substitute or explanation on how waves propagate in vacuum, the de Broglie matter waves were not a surprise. **Even if a medium was found or postulated, we still have to find out the constituents of such medium. As such, ether is as good as a field or a wave as far as the inevitability of comprehending the very essence of matter and wave.** Einstein's bent space geodesics offers nothing new other than that matter bends the space, which is, by itself, undetermined.

Louis de Broglie (15 August 1892 – 19 March 1987), France.

De Broglie obtained a relationship connecting the length of the new waves with the mass and velocity of the moving bodies as follows.

$\lambda = h / m v$

In this relation, lambda λ denotes the de Broglie wavelength, and m and v are, respectively, the mass and velocity of the body; h the Planck constant. This is significant because it means that the de Broglie waves are of a quantum nature. De Broglie's equation appears as if it was obtained by a game of crosswords. From Planck's relation we know that the photon energy is given by $E = h v$. That gives the photon a momentum $p = E / c = h / \lambda$. Equating the momentum of photon by the momentum of mass m moving at velocity v, we get $p = m v = h / \lambda$, which is De Broglie's relation. It will however be evident that, such simple juggling of variables was invaluable in defining the physical boundaries of particles and their waves. De Broglie's matter wave is extensively used in the explanation of the spatial relationships of particles and their waves during interaction with other entities.

These wavelengths should be extremely small, since the numerator is Planck's constant, which is exceedingly small: 6.6×10^{-27} erg per second. For an electron of a mass of about 10^{-27} gram and a velocity of 6×10^{7} centimeters per second, the de Broglie relation gives us

$$\lambda = \frac{6.6x10^{-27}}{6x10^{7}x10^{-27}} = 10^{-7}\,cm$$

This is something quite different as it corresponds approximately to the wavelengths of X-rays, which can be detected. Thus, in principle, we should be able to detect a de Broglie electron wave.

An attempt was made to detect the de Broglie wave in a diffraction experiment, since diffraction is so completely a wave phenomenon. Diffraction consists in the fact that when a wave encounters some obstacle it passes round it. In doing so, the wave is slightly deflected from its straight path and moves into the 'shadow' behind the obstacle. The diffraction pattern of waves from a round obstacle or a round aperture in a screen opaque to waves is typically a system of alternate dark and light rings.

It was precisely the discovery of the diffraction of light at the start of the nineteenth century that served as a most convincing argument for the wave theory of light. But the wavelengths of light waves are hundreds and even thousands of times greater than those of the de Broglie waves of electrons. All the devices constructed for producing diffraction of light-slits, screens, diffraction gratings, were much too crude. The dimensions of the obstacles used to observe diffraction of a wave must be comparable with or less than the wavelength.

By 1924, it was known what objects to use in attempts to detect the diffraction of the de Broglie electron waves. Twelve years before, the German scientist Laue had noticed the diffraction of X-rays on crystals.

Laue noticed a series of dark and light dots on a photographic plate exposed to X-rays that had passed through a crystal. Several years later, Debye and Scherer repeated Laue's experiment on small-crystal samples of powders, and obtained diffraction rings.

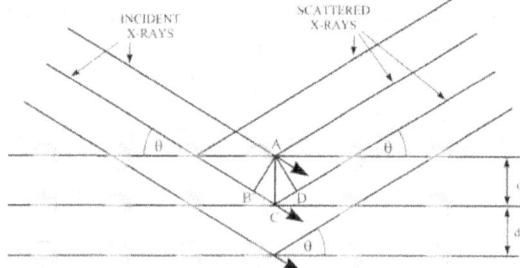

X-ray diffraction in crystals.

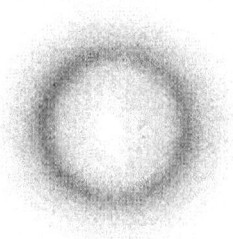

X-ray diffraction showing the dark and bright rings.

In these cases, diffraction was possible because the distances between the atoms in the crystals (like slits in a 'screen' opaque to X-rays) were of the same order of magnitude as the wavelength of the X-rays: 10^{-8} centimeter. The lengths of the de Broglie waves lie precisely within this range. Thus, if these electron waves do really exist, then electrons, in passing through a crystal, should produce the same diffraction pattern on a photographic plate as the X-rays.

A few years after de Broglie advanced his new concept, the American scientists Davisson and Germer and the Soviet physicist P. Tartakovsky verified it in an experiment on the diffraction of electrons by a crystal.

However, in itself the analogy between the electron rays' and X-rays was not enough. The experiment required great ingenuity. X-rays passed through the crystal almost unimpeded, while electrons were totally absorbed in a layer of crystal only a fraction of a millimeter thick.

What was needed, therefore, was very thin crystal plates, or metal foils, or maybe to work with obstacles and not apertures. In this ease, a beam of electrons was directed at a small angle to the face of the crystal so that the electrons sort of slid along it without going deep into the crystal and bouncing back from it. As a result, the electrons experienced diffraction only on atoms in the outermost layers of the crystal.

The electrons that had experienced diffraction were recorded on photographic plates. Tartakovsky sent a beam of electrons onto a thin foil consisting of a multitude of minute crystals. De Broglie's bold hypothesis concerning matter waves was confirmed by experiment.

In 1928, G. P. Thomson in Scotland used high speed electrons ranging from 10,000 to 50,000 volts, diffracted by very thin metallic films. His experimental arrangement is as follows

A beam of cathode rays is produced in a discharge tube by means of an induction coil. The electron pass through a diaphragm tube A to obtain a fine pencil of electrons which is then allowed to fall upon a very thin metallic film F of gold or aluminum, etc. The film should be extremely small in thickness, of the order of 10^{-6} cm, to obtain good results. P is a photographic plate which can be moved down into position to receive the pencil of electrons after it has traversed the film. S is a fluorescent screen which can be used instead of the photographic plate for visual examination of the result obtained by the passage of electrons through the foil.

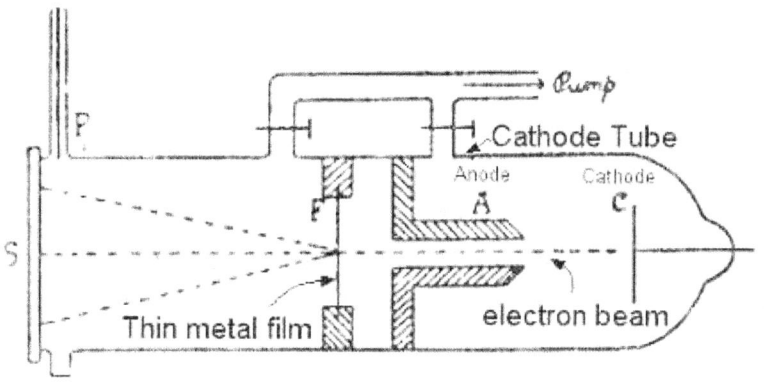

G.P. Thomson's apparatus for the diffraction of electrons.

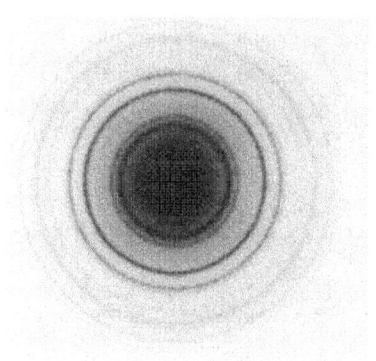

Electron diffraction pattern

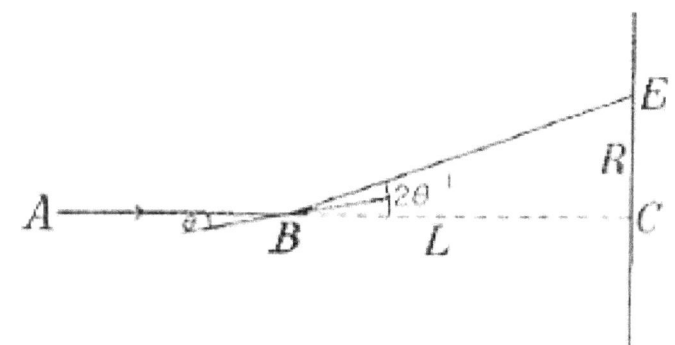

Thomson analyzed his results as follows.
Let AB be the incident beam passing through the film at B.
Let BE be a beam which has suffered a Bragg reflection in some small crystal in the film at B and falls at the point E on the photographic plate at a distance R from the central point C.
Let the distance BC from the film to the plate be = L.
The angle CBE = 2θ, where θ is given by the **Bragg's relation** $n \lambda = 2d \sin \theta$.

Now,

$R = L \tan 2\theta = L.2\theta$,

since 2θ is small. But from $n\lambda = 2d \sin\theta = 2d\theta$.

$\theta = n\lambda/2d$.

$R = L n\lambda/d$

De Broglie's value of λ for electrons of mass m and electric charge e accelerated in an electric potential difference V, in volts. is obtained from the two relationships

$\lambda = h / mv$ and $\frac{1}{2} m v^2 = eV/300$,

Substituting the values of

$h = 6.55 \times 10^{-27}$ erg/sec

$m = 9 \times 10^{-28}$ gm. and

$e = 4.77 \times 10^{-10}$ e.s.u.,

and relativistic correction α, we get

$\lambda = (150 / V)^{\frac{1}{2}} (1 + \frac{1}{2}\alpha)^{-\frac{1}{2}}$

where

$\alpha = e V / 300 \, m_o c^2$

Thus, the radius R of the diffraction ring is

$R = (nL / d) . [150 / (V (1 + \frac{1}{2}\alpha))]^{\frac{1}{2}}$

Therefore

$R V^{\frac{1}{2}} (1 + \frac{1}{2}\alpha)^{\frac{1}{2}} = (nL / d) . (150)^{\frac{1}{2}} = $ constant

William Lawrence Bragg (31 March 1890- July 1971), England

If D be the diameter of the diffraction ring considered, $R = D/2$ so that

$D V^{\frac{1}{2}} (1 + \frac{1}{2}\alpha)^{\frac{1}{2}} = $ constant

This relation has been verified by Thomson using different voltage electrons and measuring the diameter of a given ring in the pattern.
De Broglie's law was verified within the limits of experimental error.
In 1915, Bragg discovered the Bragg law of X-ray diffraction, which is basic for the determination of crystal structure.

Dual Properties Of Electrons

Even before these decisive experiments, scientists were trying to get at the real meaning of the de Broglie waves.

How was one to understand this dual nature in the behavior of particles, of electrons?
In those days, physicists knew what an electron was. A very small and very light particle of matter carrying a minute electric charge. For a long time no one asked what shape this particle had or what occurred inside it. There was no way of actually observing an electron, to say nothing of trying to figure out its internal structure.

But if an electron is a particle, then it obviously must have the properties of a particle. Electrons can fog a photographic plate in the same way that visible light of X-rays do. How could an electron have the properties of waves, something so utterly different?

Fourier waves overlapping in stationary wave packets that resembles the formation of particle and void.

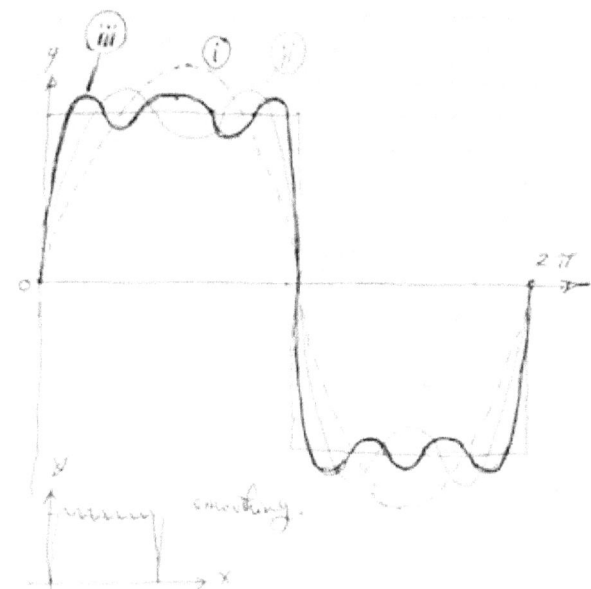

Indeed, de Broglie's matter waves were not news to mathematicians. Maxwell's wave equation treated barriers between different media transition in the gradient of the field. Thus, a particle could be accounted for in the wave theory as a sum of Fourier waves of various wavelengths that overlap in such manner that the sum vanishes outside the particle but exists as sharp barrier on the particle-field interface. In other words, a particle in the wave theory is a product of **constructive interference** of waves, while the void corresponds to **destructive interference** of those waves. As the particle moves, the interference pattern shifts its phase such that the boundaries of interference pattern define the particle. Hence, de Broglie's matter waves rejoined Schrödinger's wave equations as a special case of Maxwell's wave equations.

In the Bohr-Rutherford theory, the atom was like a planetary system in which the electron revolved about a nucleus, the only difference being that, unlike the planets, the electrons could frequently change their orbits. Then, came the light quantum, the photon. As Einstein had shown, it too possessed the properties both of waves and of particles. Now, with de Broglie's discovery, a traveling electron would, as a result of diffraction, move round an obstacle and get behind it due to de Broglie waves.

At small velocities of motion of an electron, the length of the electron wave is many thousands of times greater than the electron. As the velocity increases, the particle pulls the wave into itself, the wave becomes shorter. Even at high velocities of motion, the length of the electron wave is still greater than the 'dimensions' of the electron itself. The wave is connected with the electron intimately and for all time.

The electron wave disappears only when the electron stops. At this instant the denominator in the de Broglie relationship becomes zero and the wavelength, infinity. In other words, the crest and trough of the wave move so far apart that the electron wave ceases to be a wave. The de Broglie picture is quite vivid: an electron riding its own wave.

But where did the wave come from? It exists with the particle even when the latter is in motion in an absolute void. But again, what is that waving if there was no medium? The wave theory explains that the particle is made of the field interactions of a congregation of waves. Thus, particle waves and their particles are inseparable.

Why not imagine the wave itself to be the particle? In other words, picture the particle as a compact formation of its waves, a wave packet. A packet was to consist of a small number of rather short waves; when two or more packets collide they ought to behave like particles-exactly like a short-wave photon when it ejects an electron from a metal.

But no matter how compact the packet, no matter how much it resembles a particle, it consists of waves. This view will latter be proven wrong by Dirac who would postulate the possibility of conversion of waves to particles and vice versa. These wave packets rapidly disintegrate in time, even in a total vacuum. In negligible intervals of time, a packet becomes so clouded out in space that the formerly compact particle disintegrates into smaller proportions. The mechanical combining of two such mutually exclusive entities as waves and particles into a single image came later.

Electron Diffraction
The electron beam produced by an incandescent metallic filament was specially formed. A diaphragm with a small circular aperture was inserted between the source and the crystal. As a result, after the electron beam had passed through

the diaphragm it had definite cross-sectional dimensions. Obviously, as long as the number of electrons participating in diffraction is small, no wave properties are exhibited. These properties appear only for large numbers of electrons.

In other words, the wave properties of particles seem to be manifested only by large assemblies. A powerful source of electrons forms diffraction pattern quickly. A weak source of electrons requires long exposure time to produce diffraction. If in both cases, the same number of electrons impinges on the plate, absolutely identical diffraction patterns will be produced. Thus, each of the electrons displays its unusual properties independently of the others as if no other electrons existed at all. In other words, the electrons should reproduce on the photographic plate a whole group of light and dark rings. The distribution curve of electrons on the photographic plate is wave-like in shape. This same wave-form is exhibited by the intensity of the diffraction pattern due to light, and to X-rays, which are definitely waves.

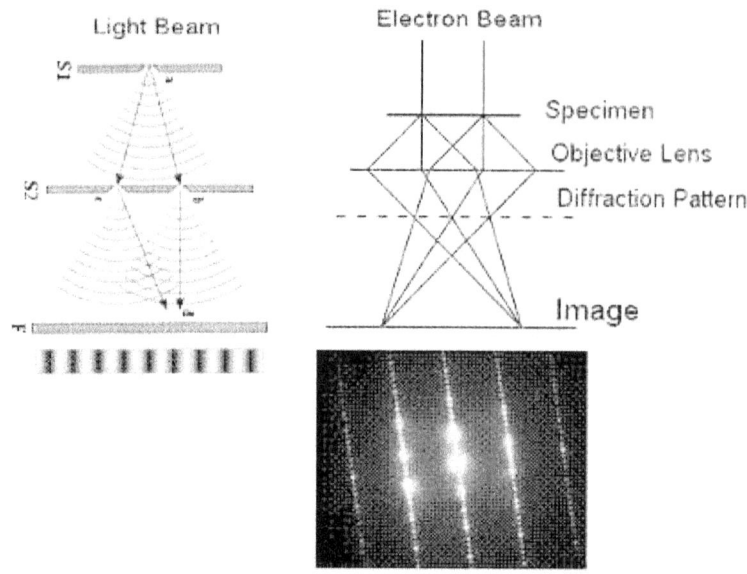

It was already known that the harder the X-rays, the shorter their wavelength and the more compressed their diffraction pattern. A study was made of the diffraction of electrons having different velocities. And here too, a tightening of the diffraction rings, as the electron velocities increase, was established quite definitely.

Now, as for the distance between the rings of the electron diffraction pattern, calculations showed that if one computed the length of electron waves from the distance between rings, the result was values that coincided exactly with those found from the de Broglie relationship.

The wavelengths of the de Broglie waves rapidly fall off with increasing mass and velocity of the particles, and lie beyond the sensitivity of our instruments. Then only the corpuscular properties of particles remain. Up to a certain point, waves (for instance, electromagnetic waives) do not exhibit any corpuscular properties and behave as waves should: they experience interference, diffraction, and so forth.

But **as soon as their wavelengths become small enough, they begin to act like particles** and are able to knock electrons out of a metal. The best example is gamma rays, the shortest of all known electromagnetic waves. They dislodge particles of matter with great ease, exhibiting true corpuscular properties.

De Broglie's discovery united into a coherent whole the world of physical phenomena, bridging the gap between two such opposites and, what would appear to be, mutually exclusive entities, as particles and waves.

Quantum Wave Theory

Bohr's description of electrons flying in privileged orbits, that do not radiate, fell on the mathematical ears of Schrödinger and Heisenberg as a very accurate description of an eigenvalue problem that could be properly formulated by a time-space differential equation. Putting the Coulomb's law of electric forces and Planck's quantization of energy into the boundary and initial conditions of the differential equation, should yield exactly the characteristics of Bohr's orbits. Further, the mathematical model would impose additional conditions on the orbits such that experimental data could be fitted properly with the solution of the equation. Those additional conditions turned out to be the spin, parity, orbital numbers, and negative total energy. Each newly imposed mathematical number led to new discoveries of elementary nuclear elements.

De Broglie's matter waves were recycled through rigorous mathematical modeling by Werner Heisenberg and Erwin Schrödinger.
Schrödinger formulated the quantum wave theory to describe electronic orbits and atomic spectra as eigenvalue problems. Schrödinger altered the Maxwell's wave equation so that the latter took into account the corpuscular 'taste' of the de Broglie waves. The new equation was called Schrödinger's equation and is the most popular equation of Quantum Mechanics. Thus, the wave law became the basic law of Quantum Mechanics.

Heisenberg and Schrödinger presented another view of the de Broglie waves. Both formulated the Quantum Wave theory that describes the motion of electrons and other particles of the subatomic world and embraces hundreds of interlinked phenomena. Mathematically, it is written in the form of the so-called wave equation and serves as the cornerstone of Quantum Mechanics. The new law of Quantum Mechanics had to be at least as general and broad such that it could describe both the motion of particles and the propagation of waves.

In 1922, the diffraction of electrons was decisively settled by experiment. In 1924, de Broglie attached the matter waves to all moving material objects. But, both de Broglie and Bohr were skimpy on mathematical formulation. All they described was the wavelength of particles, and nothing about particle motion or the intensity of radiation. In 1925, theoretical physicists had new laws far and beyond those upon which Maxwell formulated his wave equation in 1860's. Maxwell's equations yielded the strength of electric and magnetic fields as functions in time and coordinates in such manner as to conserve the charge, energy, and momentum of the system.

Erwin Schrödinger (August 12, 1887- January 4, 1961), Austria.

Werner Heisenberg (5 December 1901-1 February 1976), Germany, introduced the so-called matrix form of Quantum Mechanics.

Now, in dealing with particles, Boltzmann and Planck utilized probability functions to describe the state of particles in terms of their energies. Thus, the 1925's theoreticians must synthesize both wave and probability in their approach to the dual nature of the subatomic particles. Schrödinger and Heisenberg were the first to achieve success in this respect. Their approaches were quite different. What is more, one probably didn't even know what the other was doing. It was only some time later, after their papers were published, that Schrödinger was able to prove that both solutions of the problem were physically identical despite the fact that outwardly they had nothing in common.

Probability Wave Function

In the middle of nineteenth century physics undertook the study of internal motion in gases. It was evident almost immediately that one could not apply Newton's equations directly to the motion of gas molecules. Even small volumes of gas contain millions upon millions of millions of molecules. In order to give an accurate picture of their motion would require writing and solving the equations of motion of each of the molecules. The molecules are never at rest, they are constantly colliding with other molecules, bouncing off some, running into others; and these events occur millions of times every second.

Thus, probability was introduced to describe the state of entire mass of gas: its temperature, density, pressure and other characteristics. There is no need to determine the velocities of the separate molecules. All characteristics of the state of the gas should refer to the whole system of molecules as an assembly. These characteristics are determined mainly by the mean velocity of the gas molecules: the higher the velocity, the higher the temperature. If, in the process, the gas does not change its volume, then there will be an increase in pressure with rise in temperature. To learn these relationships accurately, one had to find some way to determine the mean velocity of the molecules. Despite the random nature of these changes in velocity, there exists, at every instant of time, some mean, stable molecular velocity under the given conditions.

Max Born (11 December 1882-5 January 1970), Germany.

What is random as concerns one molecule becomes a regularity when applied to a large number of molecules. Such is the probability law of large numbers such as the number of molecules of volumes of gas. The molecular density of gases is so large that the law can be applied without the slightest hesitation or doubt. Statistical Mechanics describes the motion of a gas by means of probability laws, by the laws of statistics, underlying them all are the exact laws of Newtonian mechanics.

In the realm of subatomic particles, new laws govern the interactions of particles differently from the large molecules of gases. Those discovered law of spin, separate energy levels, parity, quanta, and matter waves should distinctly determine the probability functions of the subatomic particles over gaseous molecules. Quantum Mechanics grew out its predecessor; Statistical Mechanics and Maxwellian Electromagnetism.

The basic theme of Quantum Mechanics was putting flesh on the skeleton, those new and old laws that govern the new world of subatomic particles. In reconciling the skeleton of the new experimental data and newly formulated laws of quanta and privileged electronic orbits, the theory offered researchers a tool measure that helps determine the congruence of ongoing research with the canons of the established laws and facts.

The Uncertainty Relation

de Broglie matter waves deflect the electron from a classical trajectory of motion. Without this deflection there would not be any diffraction pattern at all. Suppose we want to take a measurement of the velocities and positions electrons in space at every instant of time. To get clear image, the wavelength of light must be less than the dimensions of the electron. When the dimensions of the objects are of the same order of magnitude as the wavelength of light, we have a strong diffraction of the light. Inversely, with still smaller object, the light goes past it as if the object didn't exist at all.

In order to assess the interaction of photons with electrons, we apply to the photon the de Broglie relation:

$$\lambda = \frac{h}{mv}$$

where the velocity v is made equal to the velocity of light c. Then we can find the mass of the photon (this is naturally the mass of a moving photon; the rest mass of a photon is strictly equal to zero)

$$m = \frac{h}{\lambda c}$$

Now, the momentum of a photon is the product of its mass by its velocity:

$$p = mc = \frac{h}{\lambda}$$

From this formula it is readily seen that as the wavelength of the photon decreases, the momentum increases rapidly. When a light photon hits an electron, it imparts to the latter its momentum and bounces off according to the laws of reflection of light. A gamma photon with a momentum almost a thousand million times greater than a photon of light collides with an electron, the electron is knocked right out in flight; moving in some direction but we can't say with what velocity. We then illuminate it with a gamma photon and the electron changes its speed. Thus, we can't locate the electron because just as we illuminate it, the electron is knocked off again in some direction.

Even if the electron velocity is close to that of light (3×10^{10} cm/s), it will have a momentum of only about 10^{-17} g.cm/s.

The gamma photon used for illumination has a very short wavelength (say, 6×10^{-13} cm) and a momentum of 10^{-14} g.cm/s, which is thousands of times that of the electron. Thus, the possibilities of measuring instruments in the world of the subatomic particles are limited. They cannot measure particle motions with any degree of accuracy since instruments interfere with the energy and momentum of motion that are being measured.

In 1927, Heisenberg derived the uncertainty relation from the general laws of Quantum Mechanics as follows

$$\Delta x \, \Delta p \geq \frac{h}{2}$$

Or

$$\Delta x \times m\Delta v_x \geq \frac{h}{\pi}$$

Here, Δx is the uncertainty measurement of the coordinates x of a particle; Δv_x is the uncertainty measurement of its velocity v in a direction x; m is the mass of the particle, and the sign signifies that the product of these uncertainties cannot be less than the quantity on the right side of the relationship. Thus, if we try to measure the position of a particle with absolute accuracy, the uncertainty of its coordinate Δx must, of course, become zero. But then, by the rigorous laws of mathematics, the uncertainty of velocity becomes

$$\Delta v_x = \frac{h/m}{\pi \Delta x} = \frac{h/m}{0} = \infty$$

or infinity. Thus, the velocity of the particle at the instant when its position is being measured becomes indeterminate. Conversely, if at some instant of time we measure the velocity of the particle with absolute accuracy, we will have no way of saying where the particle was located at that instant. The uncertainty in the position of the electron will be thousands of millions of times the size of the electron. In other words, the uncertainty relation puts an upper limit on the accuracy of instruments. Our instrument interferes directly with the phenomenon under observation and alters its natural course. It changes it to such an extent that we have no way of separating out the phenomenon in pure form.

Again, from de Broglie matter wave, an electron at rest has its wave extending to infinity, and any attempt to find it at any definite place must fail. On the other hand, the faster an electron moves, the more accurately it is 'localized' in its wave. But even at the very highest velocities of motion its 'wavelength' is still many times greater than its own 'size'.

Tunnel Effect

The concept of potential wells and potential barriers are inherent in Quantum Mechanics by the very design of the wave equation. The pillars of invariability of total momentum and total energy introduce potential barriers to those invariables in the form of simple algebraic quantities. The plot of potential energy-coordinate, between two locations of different potentials, is called a **potential barrier**. Coulomb's potential is effected by the attraction of opposite charges. It binds the electrons to the nucleus. Inside the nucleus; nuclear forces exert potential barriers within the dimensions of the nucleus. A metal contains a multitude of almost free electrons that are relatively feebly bound to their atoms. But, despite their freedom, free electrons do not leave the metal of their own free will. They are not completely free. Though the bond is weak, the electrons are still attracted to the ions of the metal.

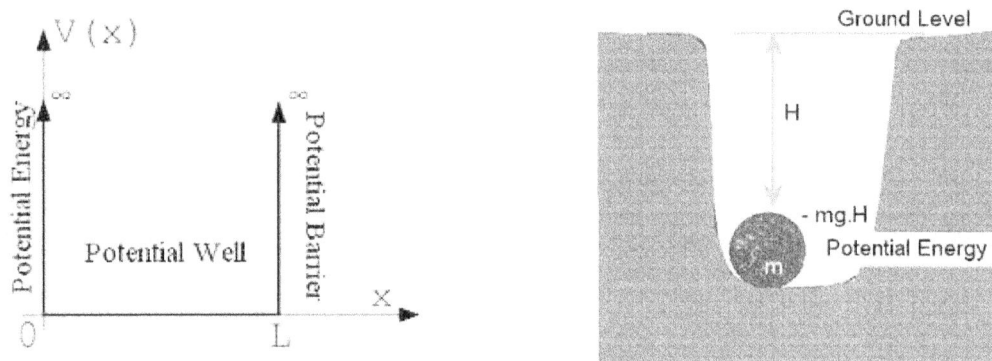

The overall action of all the ions on all the electrons in a piece of metal is restricted by potential barriers. The electrons in metal move at random, but they can no more get out unless their invariable of total energy is stepped up above the barrier by either absorption or emission of radiation. Still, these electrons are not really 'chained' to the metal for all time. Under certain conditions they can jump over the fence and get outside. For instance, this occurs when the metal is illuminated with light of sufficiently short wavelength. An energetic photon can knock an electron free, right over the potential barrier. The energy transferred from the photons of light into the bound electrons results into stepping up its energy invariable above the height of the potential well.

In the case of electrons in a metal, the barrier can also be effected by applying to the piece of metal a strong electric field.

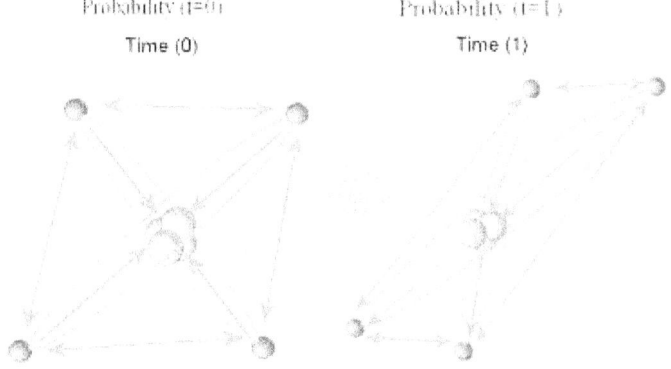

The electron in the metal is subject to constantly changing potential well. The subatomic particle is subjected to persistently changing attraction and repulsion from other electrons and from the nucleus that are persistently dependent of the emission or absorption of radiation. As the neighboring electron approach closely, their repulsion increases. As the electron moves far from the nucleus, the attraction decreases. All those motions happen simultaneously and in extremely short time and distances such that the probability of crossing over the potential barrier is completely different from a classical potential well such as the described by Bohr Atom Model. Put differently, Bohr estimated the radius, energy, and frequencies of his privileged electronic orbits with **total disregard to the temporal dynamism of neighboring electrons on each individual orbit.** Even the nucleus of the atom must bounce or orbit as the rest of electrons impart attraction and repulsion while in motion. Indeed, Bohr Atom Model froze in time and place in utter detachment from the balances of forces in the atom.

Now, if we solve the Schrödinger equation for an electron in a metal placed in an electric field, the result will be quite unexpected. Here, the probability of an electron getting out of the metal is not equal to zero and, strictly speaking, never actually becomes zero. It is very small, maybe negligibly small, but it never vanishes. Hence came the name- the tunnel effect.

From Classical Mechanics, we already know that the total energy of the ball in a hole dug in the ground is equal to the sum of its kinetic and potential energies and is negative. This is because the potential energy of the ball is reckoned from the top of the hole, that is, the ground level or the highest point of the potential barrier. The potential energy is negative and exceeds in magnitude the kinetic energy of the ball. Thus, within the limits of the barrier the total energy of the ball should remain negative, since in the process of seeping through it does not change in magnitude. But, on the other hand, the potential energy decreases until at the uppermost point of the barrier the total energy comes zero. The only conclusion is that within the limits of the barrier the kinetic energy of the ball became negative.

Schrödinger Equation
In 1860, James Maxwell collected the work of Volta, Ørsted, Faraday, and Ohm and formulated the equations of electromagnetic wave propagation so as to preserve the conservations of mass, charge, energy, and momentum in accordance with Newton's laws. Thus, the classical wave equations reflected the accumulated experimental results prior to the discovery of the discrete spectrum of elements or the infinite release of energy from radioactive material.

What purpose did the Maxwell's wave equations serve?
Well, it erected the mathematical framework that helped predict the outcome of many engineering applications. The variables of the wave equation were the electric and magnetic field, the electric charge, the coordinates and time, and the properties of the medium, receivers and transmitters. Further, the wave equation offered theoreticians a penetrating insight into the laws of conservation of the new variables of electricity and magnetism that differed significantly from the classical variables of tangible material objects.

Quantum Mechanics united two hypotheses- that of the Planck hypothesis on energy quanta and that of de Broglie on matter waves-and demonstrated their intimate interrelationship. In something like five year following the birth of de Broglie's ideas, the methods and mathematical apparatus of Quantum Mechanics were worked out in every essential detail, results of great scientific value were obtained, and far reaching attempts were made to assimilate these results.

Schrödinger in 1926 directly started with De Broglie's idea of matter waves and developed it into a rigorous mathematical theory which has received the name of Wave Mechanics. The essential feature of this theory is the incorporation of the expression for the De Broglie wavelength into the general classical wave equation. According to the De Broglie theory, a particle of mass m and moving with a velocity v has associated with it a wave system of some kind. Schrödinger traced the same path followed by Maxwell and arrived to the following equation for a free massive particle:

$$ ih\frac{\partial}{\partial t}\Psi(\mathbf{r},t) = \left(-\frac{h^2}{2m}\nabla^2 + V(\mathbf{r}) \right)\Psi(\mathbf{r},t) $$

where

$\psi(r,t)$: is the wave function; the probability amplitude for different configurations of the system at different times,

$(ih/2\pi)\ \partial/\partial t$: is the energy operator (i is the imaginary unit and $\hbar = h/2\pi$, is the reduced Planck constant),

$-(h/2\pi)^2/2m$: is the kinetic energy operator, where m is the mass of the particle.

∇ : is the Laplace operator. In three dimensions, the Laplace operator is

$V(r)$: is the time-independent potential energy at the position $\mathbf{r}$.

Like Maxwell's wave equations, Schrödinger's equation is a second order partial differential equation by the very constraints of Newton's second law of motion. It describes quantities that vary in space and in time. The unknown quantity may be the form of the surface, the coordinates of a particle, the wavelength of a wave, the speed, and many other things. The solution of such an equation yields directly the dependence of the sought quantity on other variables of interest to scientists. The unknown quantity in the Schrödinger equation is called the wave function. The square of the wave function has the meaning of probability. Its dependence on the coordinates and time yield the probability of finding a particle at some place at a given time. More precisely: the probability that a particle may be detected in a given place at some time due to the action that it performs there, for example, due to its interaction with an applied field.

Schrödinger's wave equation was expected to determine, among other properties of matter, the magnitude of the brightness of the spectral lines of different elements. Though we have no knowledge of what it is that vibrates, we can indicate it by ψ, periodic changes in a system of stationary waves to be associated with the particle and referring the particle to the Cartesian coordinate system, at any point x, y, z in the immediate vicinity of the particle, ψ, undergoes periodic changes, its value at any instant t being given by

$$\psi = \psi_0 \sin 2\pi \nu t \qquad \ldots(1)$$

where ψ_0 is the amplitude at the point considered, independent of t but a function of s, y, z, and ν the frequency. The differential equation of this wave motion can be written in the classical way as

$$\frac{\partial^2 \psi}{\partial t^2} = \nu^2 \left(\frac{\partial^2 \psi}{\partial x^2} + \frac{\partial^2 \psi}{\partial y^2} + \frac{\partial^2 \psi}{\partial z^2} \right) \qquad \ldots(2)$$

From equation (1)

$$\frac{\partial^2 \psi}{\partial t^2} = - 4\pi^2 \nu^2 \psi = - \frac{4\pi^2 \nu^2 \psi}{\lambda^2}$$

Substituting this in equation (2)

$$\nabla^2 \psi + \frac{4\pi^2}{\lambda^2} \psi = 0 \qquad \ldots(3)$$

The Wave Mechanics concept is introduced by replacing λ by $h/m\nu$ from De Broglie's theory. With this change, equation (3) becomes

$$\nabla^2 \psi + \frac{4\pi^2 m^2 \nu^2}{h^2} \psi = 0$$

Now if E is the total energy of the particle and V its potential energy, its kinetic energy is given by
$\frac{1}{2} m \nu^2 = E - V$
if this is substituted in the wave equation derived above, we have

$$\nabla^2 \psi + \frac{8\pi^2 m}{h^2} (E - V)\psi = 0 \qquad \ldots(4)$$

This is known as Schrödinger's fundamental wave equation with respect to space coordinates. It is to be noted that the time factor does not explicitly enter into this amplitude equation, as is to be expected in a **stationary wave system**. The quantity , ψ is usually referred to as a wave function. The potential energy V is, in general, a function of the coordinates. Schrödinger's equation, with respect to time can be derived for the most general form of wave motion. It takes the form:

$$\frac{ih}{2\pi} \cdot \frac{\partial \psi}{\partial t} = \frac{h^2}{8\pi^2 m} \nabla^2 \psi - V\psi$$

In the application of wave mechanics to different problems, the appropriate values for the potential energy V are substituted in the fundamental equation (4), which is then solved by a suitable mathematical device.

Schrödinger's wave equation is easier to apply on stationary structures such as nuclei, atoms, molecules, crystals, and many other things which are remarkably stable. Here, the desired wave function only oscillates about a definite 'mean' form while the form itself does not vary in time. Stationary problems do not refer to processes but to the structure of systems in which processes can occur. It is very important to know the structure, since one cannot say anything about a process unless he knows under what conditions it occurs.

In stationary system, the energy of a particle does not vary with time. Here, the uncertainty of the time of measurement does not play any part since the uncertainty principle holds that:

$$\Delta x \times m\Delta v_x \geq \frac{h}{\pi}$$

$$\frac{\Delta x}{\Delta t} \times m\Delta v_x \geq \frac{h}{\pi \Delta t}$$

$$\Delta E \geq \frac{2h}{\pi \Delta t}$$

$$\Delta E \geq \frac{2h}{\pi \Delta t} = \frac{h}{\infty} = 0$$

which means that the uncertainty in measuring energy is equal to zero. In other words, under stationary conditions, the energy of a particle is determined with absolute exactitude. In the Schrödinger equation, the magnitude of this energy is a very active participant. As long as E is positive and this, as we recall, corresponds to the free motion of a particle, Schrödinger's equation has a no vanishing solution for any values of E. And this means that the square of the wave function (the probability) is likewise nonzero for all values of E. Thus, a free particle has the right to have any energy and any velocity of motion (which, naturally, can never exceed that of light) and be located at any place in space.

In the bound state of a particle as an electron in an atom, E becomes negative. The solution of the equation changes radically. It does not vanish only for certain specific values of the energy E. These values of E are called *allowed energy levels of the particle*, or the *discreteness of energy levels*. This resembles the permitted energy levels of the Bohr model of the atom.

The electron orbits of Bohr are the energy states in which the probability of an electron being there is substantially different from zero. Bohr simply conjectured these orbits, but he was not able to prove why they should exist. It is Quantum Mechanics that slipped the foundation under this hypothesis. We should say, it was the rigorous mathematical formulation of Bohr's postulates that supported his initial inference since Quantum Mechanics cannot invent solutions without accounting for the postulates made by Planck, Einstein, Rutherford, Bohr and de Broglie. The wave equation also substantiated Bohr's second postulate concerning the quantum nature of electron jumps in atoms. As can be seen from the Schrödinger equation, an electron in an atom can exist only in states of allowed energy. When transitions are made from one state to another, the energy does not change at random but in very specific quantities. It is simply equal to the energy difference between the states of a jump or transition. This energy difference is precisely the Planck quantum.

By 1928, Quantum Mechanics was already a powerful field of physics, a fully established, mature science with a diversified approach to other physical disciplines. It took two hundred years for Classical Mechanics to reach this stage of perfection, and only five years for Quantum Mechanics. Yet, one could not dismiss the exponential growth of radical ideas that ensued due to the great accomplishments of Classical Mechanics. The invention of electrical motors, steam engines, and radio-telecommunication connected the minds of diverse and remote researchers across the globe. The father of modern physics, Rutherford, was born in New Zealand, traveled to England, and made his greatest contribution to Quantum Mechanics. It must be clearly evident that very few American names surfaced in the history of Quantum or Classical Mechanics prior to the steam age or the radio-telecommunication. Most contributors to science came from England, France, and Germany. Few came from Russia, Italy, Denmark, and Holland. Things changed in the twentieth century.

The first gain of Quantum Mechanics was the atom. The new physics, under Planck and Bohr, began with the atom. The atom was the first object of interest to Quantum Mechanics. Its first job was to reconsider the structure of the atom in its new light. Bohr introduced the concept of electron orbits. Quantum Mechanics replaced the orbits with distributions of the probability of an electron being located at some place in the atom, clouded with de Broglie matter wave and not a point in space. An integral number of the electron de Broglie waves fit exactly onto each one orbit or $2\pi r = n\lambda$. The first orbit, closest to the nucleus, accommodates one wave, the second, two waves, the third, three waves, and so forth. This serves as new proof of the universality of the de Broglie waves. The probability function or cloud is denser where there is more probability of the electron existing, and more rarefied or transparent where the probability is lower.

Their interaction is that of two charged particles of opposite sign but of the same magnitude. This interaction is described by the Coulomb law. The energy of this interaction is dependent solely on the distance between the electron and the nucleus. That renders the electron cloud of the Hydrogen atom spherical in shape. All points of the surface of a sphere are at the same distance from the centre, which in our case is the nucleus. Therefore, all points of the electron cloud correspond to the same electron energy.

But when there are more electrons in the atom, the picture of electrical interactions among them and also with the nucleus ceases to be so primitive, as in the Hydrogen atom. The electrons now are not only attracted to the nucleus but are repulsed by each other. The quantum wave equation was formulated to tackle such exact problem of possible arrangement of multielectrons in complex atoms such that the atom remains stable.

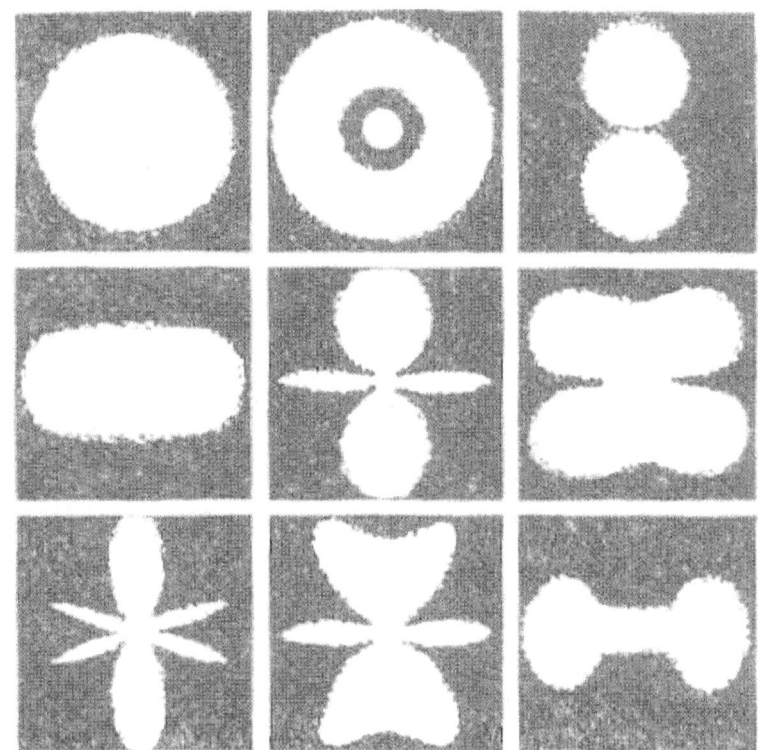

These are the density distribution of clouds of probability' of atomic electrons.

Harmonic Oscillator
As with Bohr's Atom Model, the harmonic oscillator is a perfect example that could demonstrate the application of Schrödinger's wave equation.

The equation of motion of a linear harmonic oscillator is

$$x = A \sin 2\pi\nu t \qquad \frac{d^2x}{dt^2} = -4\pi^2\nu^2 x$$

$$\text{Restoring force} = - \; m\frac{d^2x}{dt^2} = 4\pi^2\nu^2mx$$

$$\text{Potential energy } V = \int_0^x 4\pi^2\nu^2mx\,dx = 2\pi^2\nu^2mx^2$$

Substituting this value of V in Schrödinger's wave equation

$$\frac{\partial^2\psi}{\partial x^2} + \frac{8\pi^2m}{h^2}(E - 2\pi^2\nu^2mx^2)\psi = 0.$$

Using a new variable k, where k is given by

$$k = x\sqrt{b}$$

and

$$b = 4\pi^2m\nu/h$$

$$\therefore \qquad \frac{\partial^2\psi}{\partial x^2} = \frac{\partial}{\partial k}\cdot\frac{\partial\psi}{\partial k}\cdot\left(\frac{\partial k}{\partial x}\right)^2 = \frac{\partial^2\psi}{\partial k^2}\cdot\left(\frac{\partial k}{\partial x}\right)^2$$

$$\text{Since } k = x\sqrt{b}, \; \frac{\partial k}{\partial x} = \sqrt{b} \text{ or } \left(\frac{\partial k}{\partial x}\right)^2 = b$$

$$\frac{\partial^2\psi}{\partial x^2} = b\cdot\frac{\partial^2\psi}{\partial k^2}$$

$$\therefore \qquad b\frac{\partial^2\psi}{\partial k^2} + \frac{8\pi^2m}{h^2}\left(E - 2\pi^2\nu^2m\cdot\frac{k^2}{b}\right)\psi = 0$$

$$\text{Putting } \frac{8\pi^2mE}{h^2} = a, \text{ we have}$$

$$\frac{\partial^2\psi}{\partial k^2} + \left(\frac{a}{b} - k^2\right)\psi = 0$$

On the assumption that ψ, should be finite, continuous and single valued at $k = 0$, the proper solutions of this equation can be shown to be as follows:

$$\psi = \tfrac{1}{2} e^{-k^2/2} \qquad \text{when } \frac{a}{b} = 1$$

$$\psi = 2k\, e^{-k^2/2} \qquad \text{when } \frac{a}{b} = 3$$

$$\psi = (4k^2 - 2)\, e^{-k^2/2} \qquad \text{when } \frac{a}{b} = 5$$

$$\psi = (8k^2 - 12k)\, e^{-k^2/2} \qquad \text{when } \frac{a}{b} = 7$$

Thus the proper values of a/b are given by (2n + 1), where n is an integer and the values of ψ corresponding to these are known as the characteristic eigen-functions. Substituting for a/b,

$$\frac{a}{b} = \frac{8\pi^2 m E}{h^2} \times \frac{h}{4\pi^2 m \nu} = \frac{2E}{h\nu} = 2n + 1$$

$$\therefore \qquad E = (n + \tfrac{1}{2}) h\nu$$

These values of the total energy represent the quantum states of the harmonic oscillator. Thus, wave mechanics leads to the same result as quantum mechanics with reference to the energy states of a linear harmonic oscillator and *zero-point energy* which is not zero but $h\nu/2$. The characteristic functions for which n = 1, 2, 3 in the case of a harmonic oscillator are as follows.

The X-axis represents the variable *k* which is a function of *x*, while the Y-axis the normalized wave function. It can be seen that the wave amplitudes become zero even for small values of *x*. This means that the associated wave packet is found only in the immediate neighborhood of the particle, in accordance with the initial assumptions; the wave amplitudes have a finite and rather indefinite extent in space, which indicates that the *wave packet behaves very much like a particle with* fuzzy *edges.*

Pauli's Exclusion Principle
Schrödinger wave equation threw many constants when the boundary values are satisfied. Those new constants must find experimental interpretation as follows.

Atoms are made up of electrons and nuclei. The cement that keeps them together is the attractive force of the electrons to the oppositely charged nucleus. To date, 104 chemical elements are known. These are the blueprints of the structures of the universe. They look different only at first glance. The basic principle of subatomic particles in the atom was discovered by the Austrian scientist Wolfgang Pauli during the formative years of Quantum Mechanics and was named in his honor. It states: **in any assembly of elementary particles, each state of allowed energy can be occupied by no more than one particle.** Pauli's Exclusion Principle accounted for one of the constants required for the solution of

Schrödinger wave equation. There will be more new constants to account for and more new principles to interpret them and even newer wave equation will be proposed by Dirac that requires newer constants to be interpreted.

It was found later on that this principle is not absolutely universal, and that for certain types of elementary particles it does not hold. As **for electrons in the atom, no matter what kind of assemblies they form, the law never breaks down.** Thus, all the atoms of a given chemical element the electron families are absolutely identical.

The Spin
The spin corresponds to some sort of motion of the electron when the velocity is close to that of light. That occurs in atoms in excited states. It exists irrespective of whether the electron is moving fast or slow or is at rest. The value of the spin is always the same. It is an *intrinsic a property of particles* as their rest-energy. If we change the spin, we change the type of particle.

Pauli, Wolfgang (1900-1958)

Arnold Johannes Wilhelm Sommerfeld (5 December 1868 – 26 April 1951), Germany, introduced the 3rd quantum number (azimuthal quantum number) and the 4th quantum number (**spin quantum number**). In 1916, Sommerfeld advanced a more exact theory of the origin of atomic spectra to account for the newly discovered magnetic and electrical properties of atoms and molecules.

Walther Ludwig Julius Kossel (4 January 1888 – 22 May 1956), Germany, in 1914, Kossel laid the foundations of quantum chemistry, known for his theory of the chemical bond (ionic bond/octet rule), Sommerfeld–Kossel displacement law of atomic spectra.

As far back as the end of nineteenth century, it was found that if a substance is introduced into a magnetic field, its spectral lines split into various numbers of fainter lines. It was later established that a similar splitting is experienced by the spectral lines of the atoms of all elements. An understanding of the nature of this phenomenon (called the **Zeeman effect**) and especially the reason for line splitting into different numbers of 'satellite' lines was obtained only in 1925 when two young physicists Uhlenbeck and Goudsmit introduced the notion of spin.

George E. Uhlenbeck 1900, Netherlands -1988, US   Samuel Goudsmit [1902-1978]

together, discovered electron spin in 1925. This discovery was a significant contribution to the theory of theoretical physics.

The spin is incomprehensible from the standpoint of Classical Mechanics. The discoverers of spin believed that it meant the rotation proper of the electron as the earth revolves about the sun and it also rotates' on its axis. The electron revolves round the nucleus and but we have no way of sensing electron rotation about its axis. Quantum Mechanics views the electron not as a sphere but as a point. The 'axis' of a point has no meaning. But, Quantum Mechanics is not a perfect science and our sense of the spin of the electron around it axis is no different from our sense of other properties of the subatomic particles. The existence of spin in an atomic electron makes itself known from the angular momentum of the electron during its motion in the atom about the atomic nucleus. The spin of an electron is always the same and is always associated with the electron. It appears that the spin of an electron in an atom can either be added to the angular momentum of the electron or subtracted from it.

Both values of the total angular momentum of an electron correspond to opposite proper motions of the electron, which actually do not in any way differ one from the other electrons. Both these motions have the same energy in an atom not affected by any external forces. **As a result, each level of allowed energy in the atom may be occupied by two electrons instead of one, with spins directed in opposite senses.**

Electronic Architecture Of The Atom
The mutual repulsion of electrons should greatly increase the potential energy of an atom. A structure is more stable if the potential energy is low. In working out the atomic blueprints, the mutual antipathies of the electrons are managed through the solution of governing differential equations. Those are founded on two basic principles of the structure and filling-in of atomic buildings: **the Pauli principle** and **the principle of best energy distribution**.

The principle of the best energy distribution allows an electron to reside in the next story while the lower story is not yet completely filled such that an atom becomes more stable. The potential energy of repulsion of the electrons in such an atom is low. The following electronic structure of elements demonstrates the working of the best distribution of energy.

1. <u>The Hydrogen atom</u>

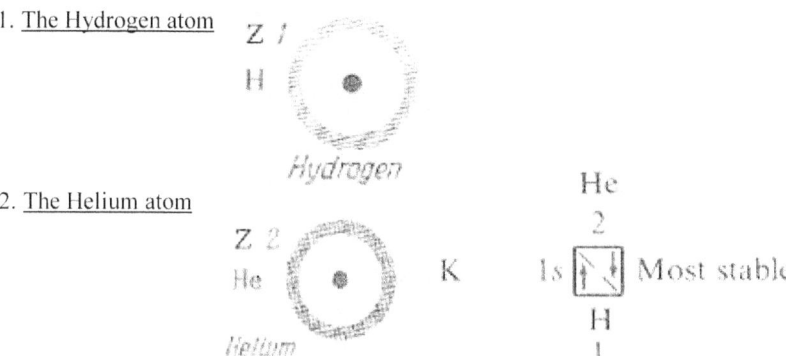

2. <u>The Helium atom</u>

Each electron cloud may be formed by two electrons. This would suggest that the Helium atom is not very different from the Hydrogen atom. The electron cloud is twice as dense and closer to the nucleus.

3. <u>The Lithium atom</u>

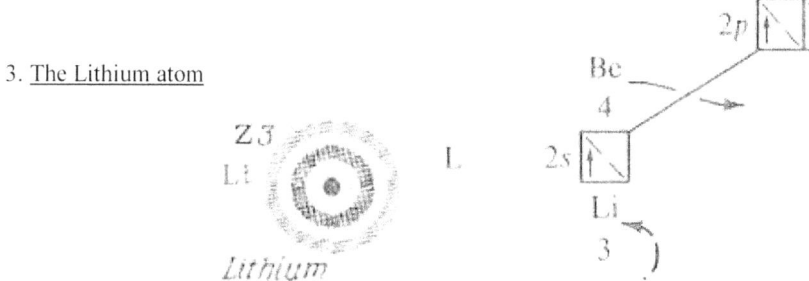

There is a second spherical electron cloud containing within it the first one of Helium. Which is natural because the Pauli principle does not allow more than two electrons in one atomic energy 'flat'.

4. <u>The Beryllium atom</u>

Beryllium is the second tenant in the second-story flat appears in the atom following Lithium So far the atomic house is filling up in orderly fashion.

5. <u>The Boron atom</u>

The new electron is placed so that it would not come into contact very often with the other tenants to avoid mutual repulsion. They all try to move in opposite directions. The new electron is placed in a separate flat right through all the stories of the atomic house.

6. The Carbon atom

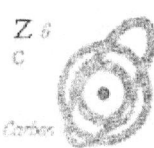

The Carbon electron will be added next to this vertical-like flat.

The de Broglie waves in an atom determine the 'capacity' of atomic structures. An electron cloud accommodates a whole number of de Broglie waves. This number determines not only the 'number' of the earlier orbit but also the density of the electron cloud formed by all the electrons whose clouds accommodate one and the same number of their de Broglie waves. This 'united' electron cloud which consists of a number of pairs of clouds is given the name *'shell'*.

Quantum Mechanics established also the relationship between the capacity of the shell (that is, the largest possible number of electrons that it can accommodate, N) and the serial number of the shell, which is the number of electron waves that fit into it, n. This relationship has a very simple form:

$$N = 2n^2$$

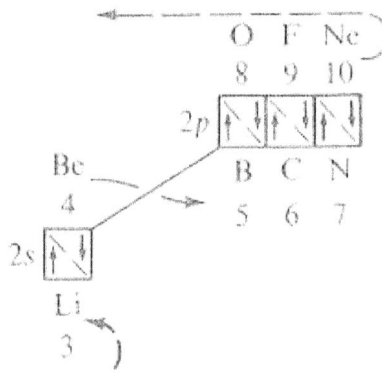

Accordingly, the first shell (K) accommodates $2 \times 1^2 = 2$ electrons.

The second shell (L) has $2 \times 2^2 = 8$ electrons.

The third (M) has $2 \times 3^2 = 18$ electrons.

The fourth (N) has $2 \times 4^2 = 32$ electrons.

The fifth (0) has $2 \times 5^2 = 50$ electrons, and so forth.

We see that the first shell to fill up is the smallest shell of lowest capacity, the K-shell. It is already completely filled in the atom of Helium. This shell is one story, with one flat for two.

The next shell is more complicated. It occupies not only the second story, but has three interstory flats, each with two apiece.

7. The Neon atom

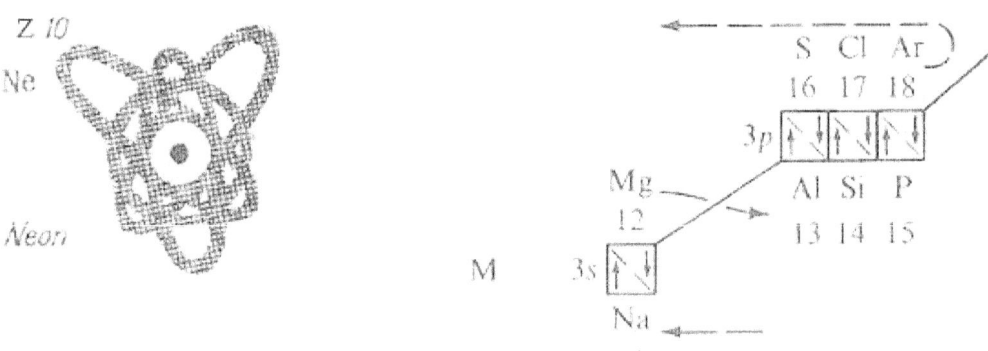

The end of the filling comes in the atom of Neon (Z. 10), which occupies the tenth square in the Periodic Table of elements.

8. The Argon atom

The third shell accommodates 18 occupants. It fills up to Argon (Z. 18) in the same fashion as the preceding shell. The first to fill in is the third story, and then three interstory flats.

But in the element after Argon, Potassium, this strict order breaks down, Here, five interstory flats have to be filled but the layout is different. Unlike the first three, these are narrower and more elongated. The new tenant goes in a new flat on the fourth floor. Another electron is added in the next atom of Calcium.

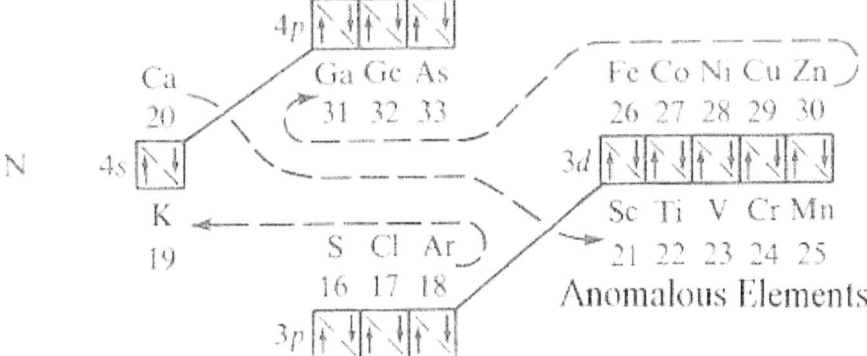

Then, the principle of best energy distribution breaks down. In nine atoms, from Scandium (Z. 21) to Copper (Z. 29), the new tenants are stuck into those long, narrow and uncomfortable flats. These atoms with flats being taken upstairs and empty flats downstairs have acquired a number of unusual properties. They are called `anomalous', and will be putting in an appearance now and then.

Strictly speaking, the third shell should fill up completely in the case of Nickel (Z. 28). But due to the violation of the principle of best distribution of energy, the third shell is completely filled only in the case of Zinc (Z. 30).

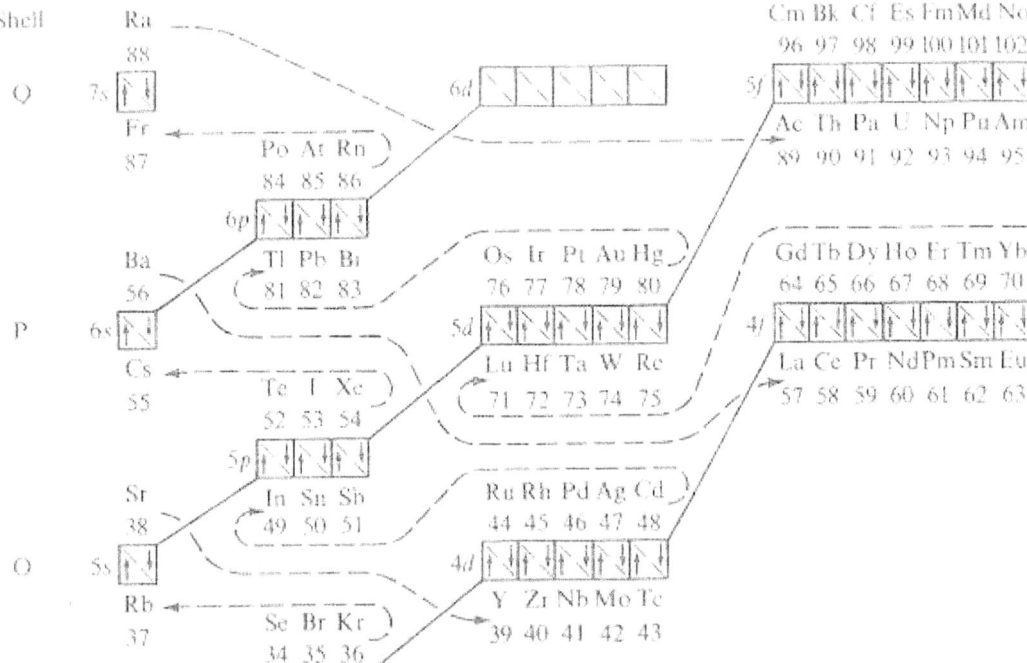

There is no improvement later on either. A shell doesn't get filled up completely before the succeeding one starts up.

What happened in the Scandium to Copper group is repeated in the group of atoms from Yttrium (Z. 39) to Palladium (Z. 46), and from Lanthanum (Z. 57) to Ytterbium (Z. 70). From then on, all the atoms right up to the last one (104) have defective tenancy rules, where even two or three shells await lodgers.

There may seem to be some lack of symmetry here, but energy-wise it is the best way to attain stable electronic structure. Thus it appears that the wave law which determines the population and order of settling atomic structures is not all-powerful. This law is frequently modified by a no less important and powerful law of the stability of atomic structures.

Periodic Table of Elements
The periodic table of elements was proposed as a simple scheme to group elements in periods of their electronic structures such that their chemical properties could be categorized in feasible manner. On the left of the table, there are seven periods, 2 in the first, 8 in the second, 8 again in the third, 18 in the fourth and fifth, 32 in the sixth (these include the rare earths, the lanthanides, at the bottom of the Table), and 17 in the seventh.

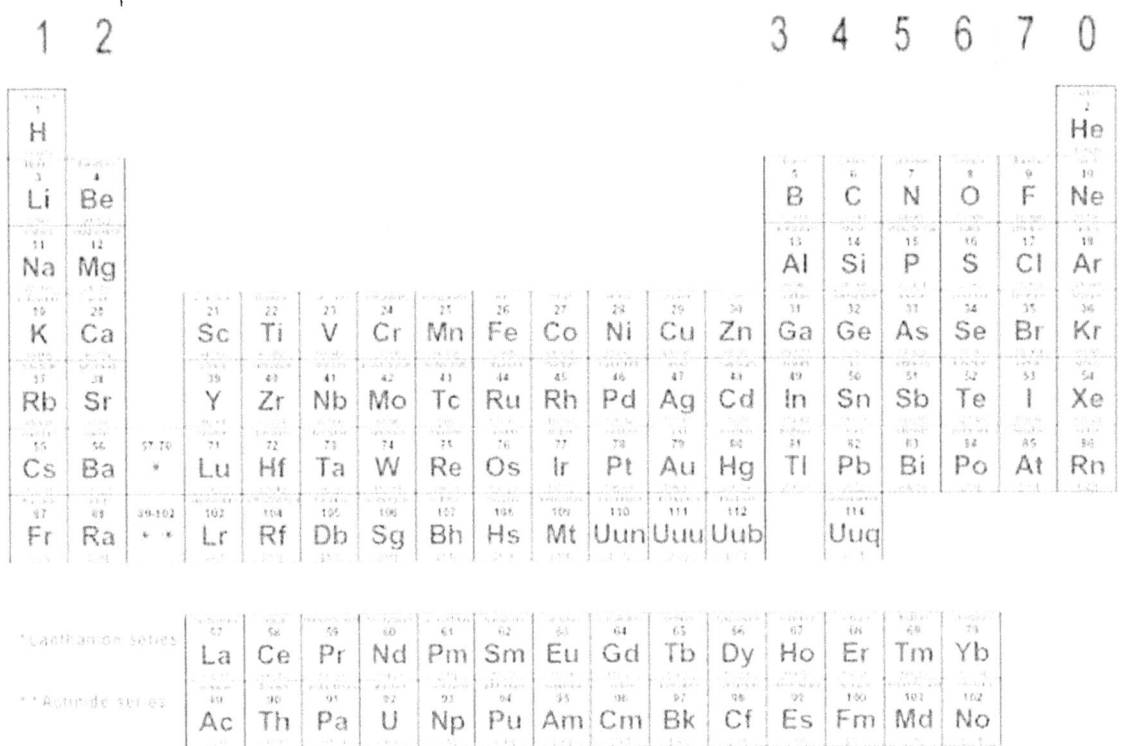

Now let us go back to the figures that describe the capacity of the electron shells: 2, 8, 18, 32, etc. They are the same as in the Table. But then why are some of the numbers repeated: 2, 8, 8, 18, 18, 32 (for the time being we disregard the last period). These repeating numbers, it turns out, are the result of those very breaks in the order in which the electrons fill in the atom. And so the third period ends with argon (Z. 18) instead of nickel (Z. 28).

From then on, this shift (and other shifts due to violations of the filling sequence) continues to the very end of the Periodic Table. As a result, we don't get an exact and simple correspondence between the shells and the periods. But the capacity of any period does not exceed the capacity of its corresponding shells. Thus the quantum picture gives a good account of one important feature of the Periodic System of elements.

Now take a look at the top of the Table. Here we have Groups from I to VIII and then 0. These are not simply valences, but valences with respect to Fluorine (or, with respect to Hydrogen). Valence with respect to Fluorine represents the number of electrons in the outermost shell of the atom, the one farthest from the nucleus. In this definition, valence coincides with the number of the group, with the exception of the last two columns of the Periodic Table. It would be more correct to put VIII over the last column and the numbers 0, 1 and 2 over the second to the last one. There are some very solid grounds for doing this.

Why does the outermost shell of an atom never have more than eight electrons?
This immediately becomes clear when we recall the order of distribution of 'living space' in atoms. The first shell accommodates only 2 electrons, the second, 8; the third should have 18, but the build-up stops for a time at Argon when there are 8 electrons. After this the outermost shell became number four, and the third shell (now an inner one) begins to fill up. The same happens to the fourth shell, and so on. As soon as the outer shell has accommodated eight electrons, any further addition of electrons is disadvantageous. But then a new shell appears and the unfilled one goes deeper into the atom.

Inner shells are *irrelevant* in chemical reactions whether they are fully filled or not. Only the outermost shell of the atom determines its chemical properties. Thus, **there are eight possible types of chemical behavior of atoms in accord with the number of electrons in their outer shell.**

Before going any further, it is necessary to point out that a completely filled eight-electron shell has a much smaller potential energy than if it had empty or partially filled 'flats' in it. Atoms with full outer complements of electrons in their shells are **'noble'** to such an extent that they eschew all contacts with the ordinary rank and file atoms. Hence their name: noble, or inert. They make up the last column in the Periodic Table.

The *'common'* atoms (those with partially filled outer compartments) experience the action of unstable forces of attraction that tend fill up their outermost shell to the eight-electron set. This process is named **'reaction'** by which one atom gives up its outer electrons to the other and remains ionized of stripped from its outer electronic coat. This refers only to atoms that come after Neon.

By way of illustration, let us take a reaction between Sodium and Chlorine, which leads to the formation of Sodium Chloride, a molecule of NaCl. The Sodium atom has an extra single electron in its third shell. The Chlorine atom has a set of seven electrons in its outer shell. The Sodium sacrifices its sole electron and Chlorine then achieves a 'noble' eight-electron outer shell. But Sodium has gained as well. It now exhibits a full complement of eight electrons of the noble gas Neon. Thus, two of the 'common atoms' become two 'nobles' at one shot, but only if they go together as a single molecule.

Atoms can, as we see, be divided into *givers* and *takers*. **Those with less than four electrons in their outer shell are givers. Those with more than four are takers.** Naturally, it is easier to acquire two electrons, say, than to give up six such as in the situation of the Oxygen atom.

In Group IV, we find a number of elements with four electrons in the outer shell. They falter in exchanging their electronic coat. Those got the name **amphoteric,** which means partly one and partly the other. These elements are capable of almost any kind of chemical behavior.

In Group VIII, we find the so called **'crazy' atoms**. They have only one or two electrons at the most in their outer shell. They come in Group VIII simply because their highest valence with respect to Oxygen can be 8, which means that each such atom can attach four Oxygen atoms to itself. Here, the underlying shell exerts a substantial and extremely complex effect on the behavior of the electrons of the outer shell. These atoms are capable of the most unpredictable things. For instance, they have a variable valence, one in one reaction and quite a different one in another.

Further, many elements in other squares of the Periodic Table exhibit similar properties like the so called 'crazy' atoms. The tricks they play are just like those of their companions from Group VIII. Periodic Table of elements does not reflect this, and one shouldn't expect it to, for the Table was devised at a time when nobody knew how the atom was constructed. There is still a great deal that we do not understand in the behavior of the anomalous elements.

Atomic Spectrum
Bohr's theory explained the origin of atomic spectra but could not give a correct description of spectral laws. It was the job of Quantum Mechanics to fill out the picture in detail. In accounting for the origin of spectra, Quantum Mechanics fundamentally agrees with Bohr's theory. The agreement was expected from the solution of the boundary-value problem

formulated by Schrödinger's wave equation where photons were given quantized energy and de Broglie's matter wave specified the distance of the electronic shells from the nucleus.

During the jumps of atomic electrons from one energy state to another, the difference of these energies is embodied as a quantum of electromagnetic energy, the photon. If the energy diminishes, a photon is born. If the energy increases, a photon or a quantum of energy of any other field has been absorbed just before the jump. The emission or absorption of a photon shakes up the atomic 'jelly', producing a new overall form. This is clearly the longest known transmutation of wave and matter that man has witnessed in the fire, light, and elements of earth.

Quantum Mechanics accounted for the change of energy state by change in their probability function. In Bohr's theory, an electron jump from orbit to orbit is always possible, and the probability of such a jump is in no way dependent on the kind of orbit. Quantum Mechanics refined such postulate and demonstrated that electron jumps have a probability that is very appreciably dependent on the shape of the electron clouds that correspond to the electron prior to and after the jump. In this situation, the probability of a jump is, roughly speaking, greater for a stronger overlapping or a deeper penetration of these clouds among shells.

The laws that divide electron transitions in atoms into more probable and less probable have become known in Quantum Mechanics as *selection rules*. The selection rules are more or less carefully observed in the light atoms where there are few electrons, so that their clouds intersect rather infrequently. But in the heavy multielectron atoms with their complex probability structures, the restrictions or prohibitions of Quantum Mechanics largely break down.

It is in this jumping of electrons in the rapidly changing tremors of electron clouds that photons are born. The photons enter a spectroscope, get sorted out into types and produce the spectral lines of all the colors of the rainbow. The more photons an atom emits in a second, the brighter the lines. And if the number of atoms remains constant, the brightness of the spectral lines can depend only on one thing-the frequency of electron jumps in atoms.

And this frequency, as we already know, is determined by the probability of jumps. Different clouds have different probabilities, some greater, some practically nil. To every photon energy and spectral line there corresponds a probability and brightness. Thus, an atomic spectrum consisting of a number of lines of different brightness is generated.

Does a photon correspond to a line of one frequency, to one wavelength? Yes.
Why do the lines on the photographic plate of a spectroscope come out rather broadened and not slender?
This was due to the wave properties of the electron with their constant attribute, the uncertainty relations. We have already said that an electron in an atom has a very definite energy. So where do the uncertainties come in? The initial energy is definite, the final energy is also definite; their difference, which corresponds to the energy of the photon, must also be an exact quantity. But, an electron jump is a violation of some steady state that is governed by the Heisenberg uncertainty principle, which in turn is founded on de Broglie's matter waves.

What is the lifetime of an electron between jumps?

That is described by the uncertainty relation between the duration of jump Δt and the photon energy as follows.

$\Delta E = h / \Delta t$

This in turn determines the frequency of the photon through Planck's relationship as follows.

$\Delta \omega = h / \Delta t$

In other words, the steadier the life of an electron in an atom, the narrower the spectral lines since these lines refer to transitions to other states, and vice versa. Further, a high temperatures and pressures, the spectral lines broaden out and become blurred.

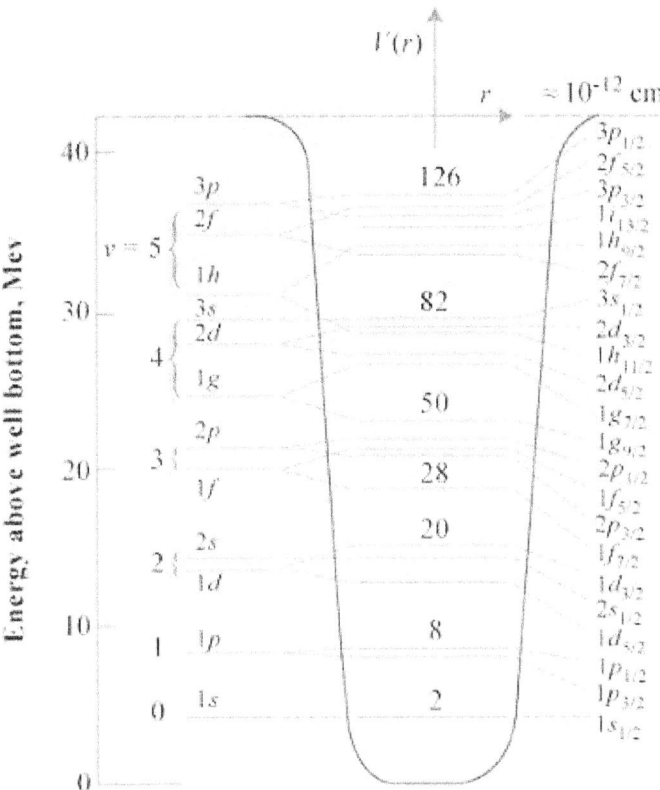

With the advance in spectral techniques, the fine structure of spectral lines revealed that many spectral lines, which would seem correspond to a single wavelength, actually turned out to be the states of a number of very closely lying lines. So, electron jumps between the same states could give rise to photons with different or even ever so slightly different energies. Spin was discovered precisely because of these 'fine qualities' in spectra. It turns out that when spectra are generated, the common state of two electrons with opposite spins has different angular momentum and thus slightly different energies, hence the doubling of the spectral lines: in place of one line we have twin lines with identical brightnesses.

CHAPTER 3:
MOLECULES AND SOLIDS

Molecules

Molecules form by the sharing of the outer electronic coat between atoms. At times, three, four and even more partners participated. A whole molecule acts closely as an inert element by imitating noble electronic coat. The molecule contains negative and positive ions. Those are the takers and givers of electronic coats. In the redistribution of electrons, the taker atom won't let go of excess electrons. The giver atom companion is subjected to unbalanced Coulomb forces. This electronic charge tie-up goes by the scientific term "**ionic molecule**".

The adhesive forces in such molecules are mainly the forces of ordinary electric attraction between ions with different charges. There is a great diversity of ionic molecules. Here, atoms from the left-hand side of the Periodic Table form molecules with atoms from the right-hand side. The farther away they are in the Table, the more closely knit is the family. The close-lying groups of atoms in the Table bond less strongly.

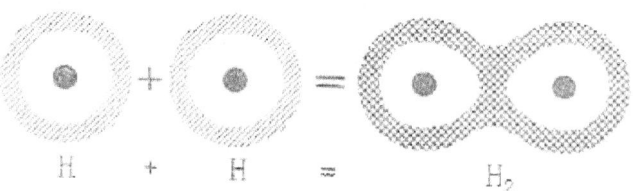

But there is just as large a group of molecules whose atoms bond for quite different reasons. The simplest family of this kind is the Hydrogen molecule. In this class of molecules, come all the single-element molecules such as molecules of Oxygen, Nitrogen, Chlorine, and also the molecules whose atoms all belong either to the left-hand side or the right-hand side of Periodic Table. These molecules came to be known as **covalent molecules**.

Covalent atoms do not have enough electrons in their outer shell. So, they share the full outer electron on time basis. One atom gives up electrons and the other acquires electrons so that the atoms in it are almost all the time ionized. First one will be surrounded by electrons and the other will be naked, then the other way around. Such process of formation of molecules is termed *'exchange interaction'*. As long as the atoms are some distance apart, their electron clouds hardly at all overlap. But when these atoms come close enough, the considerable mutual interpenetration of the electron clouds make perceptible the probability of an electron of each of the atoms finding itself near the nucleus of the partner atom, which amounts to a probability of exchange. What is this probability? About 15 per cent for the Hydrogen molecule. Thus, during 10 minutes of each hour both electrons come together in a single atom of Hydrogen, while the other atom has none. Such estimate of the duration of electronic exchange is reckoned from the solution of the Quantum Wave Equation.

The Nitrogen atom (Z.7) has only seven electrons. Two in the inner shell do not participate in exchange actions. But in the outer shells, 5 electrons exchange with other atoms. The next atom following Nitrogen is Oxygen, which has six electrons for exchange purposes in each atom. These form a molecule of ordinary two atomic Oxygen. The electronic clouds of the outer shell forms something like a figure eight.

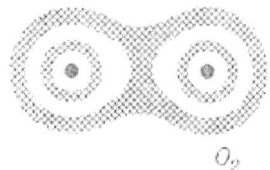

In three-atomic Oxygen or Ozone, a union of 18 electrons is formed. To simplify the exchange, the three form a triangle to cut down the distance the electrons have to cover. These three atoms toss their electrons about like a ring of players in volleyball practice. **This molecular structure no longer resembles the architecture of the constituent atoms.** And the flats together with the distribution of tenants differ too. All of which makes the properties of molecules quite different from those of the atoms they are made up of.

Quantum Mechanics of Solids

We know that solids come in crystalline and amorphous form, that they conduct heat and electricity in a variety of ways, and the transmission of light and sound is different too. Accounting for properties of solid materials is paramount to the rapidly advancing technologies and for making use of new natural materials. The demands are so great that artificial materials are pressed into service. We need materials that possess great hardness, electrical conductivity, heat resistance and many other properties. Where do we get them? Quantum Mechanics presents a versatile mathematical model for predicting the properties of solids that have not even been invented yet.

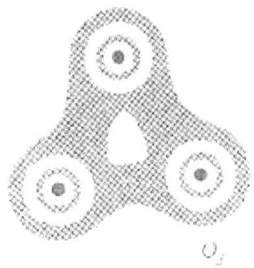

In the beginning, the attempt was made to comprehend the structure of crystals of metals. A crystal is an ordered periodic distribution of atoms in space in a lattice-like configuration. Unlike the ordinary lattice with its two dimensions, this one has three dimensions. In a lattice, the atoms of the crystal are located at constant distances, called *lattice spacing*. In the general case, there are three spacings in accord with the three dimensions of the lattice: length, width and height. Pure elements are not common in nature.

Compound crystals have lattices made up of several types of atoms. A simple case is the ice crystal which has Hydrogen and Oxygen atoms. In accord with the formula for water, the number of Hydrogen atoms is twice that of Oxygen atoms. Another case is the lattice of crystals of Sodium Chloride, NaCl. At the intersections of the elements of

the lattice, called *nodes*, we find ions of Sodium that alternate with ions of Chlorine. Note that these are ions and not atoms. It is very important that when molecules of salt 'freeze into' a solid body, the ionic nature of their atomic bonds is retained. But as such the molecule ceases to exist. It cannot be isolated. Indeed, each Sodium ion is surrounded by ions of Chlorine, and each Chlorine ion is surrounded by Sodium ions.

In a crystal like this, the forces acting between ions are ordinary electrical forces. A Sodium ion attracts Chlorine ions in the immediate vicinity, those in turn attract other Sodium ions, but repulse adjacent Chlorine ions. The interplay of forces of attraction and repulsion creates equilibrium in the ion configuration. This is the crystal lattice. If one ion gets knocked out of position, its attractive force towards ions of a different kind is diminished, but its own ions repulse it more strongly. The combined action of these forces compels the ion to return to its original position. Strictly speaking, an ion is all the time in oscillation about its stable position due to the random knocks of thermal motion, like a sphere attached to a system of springs.

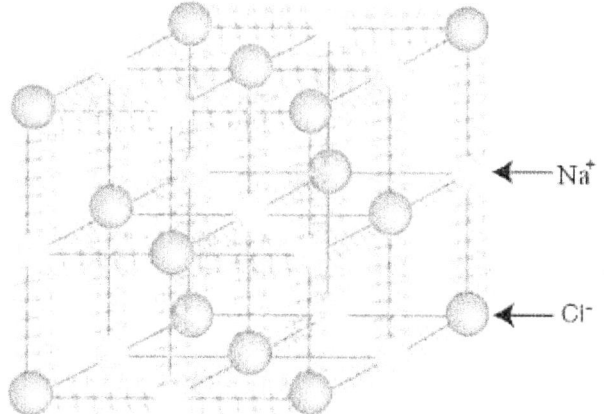

In **metallic crystals**, the most important in modern technology, the situation is quite different. Suppose that the entire lattice is made up of a single metal, that is, of atoms of one kind. Quite naturally there will be no difference in the charge of the ions. If one atom readily gives up an electron, why shouldn't all the rest do the same? The metallic crystal is indeed a gigantic *covalent molecule* consisting of many billions upon billions of atoms. Electron exchange here resembles that of the Hydrogen molecule. When the atoms of metals join to form crystals, they do actually make their outermost valence electrons common to all. This results in a sort of skeleton architecture of the crystals. At the lattice nodes are slow-moving ions surrounded by a light and mobile common cloud of electrons. This cloud plays the part of cement

The thermal vibrations of ions in a lattice determine many important properties of solids.

holding together the similarly charged ions. In turn, the ions are the adhesive that keeps the electrons from flying off in all directions. Since every atom makes its contribution to the common community, each electron ceases to belong to some one atom and is simply one of the billions upon billions of other free electrons in the crystal.

Not all electrons are as free since each of the atoms gives up only one or two of its outermost electrons, the rest are held firmly in place within the atom. Even so the army of free electrons is colossal: 10^{22} to 10^{23} in every cubic centimeter of metal. Thus, a metallic crystal has a better 'social' organization than an **ionic crystal**, where we find something like tight bonding with all the electrons chained in their atoms. This property gives the metal the opportunity to *conduct electric current*. If an ordinary electric field is applied to an ionic crystal, there will only be a slight redistribution with the electron clouds in their atoms somewhat elongated. This will result in what is known as *electric polarization* of the crystal.

Not a single electron will get away from its ion, and the ions themselves will, as before, remain firmly anchored at their nodes. And since there are no free carriers of charge, there will be no electric current. Ionic crystals are insulators.

What kind of energies do the 'collective-like' electrons in the metal have?
The answer is simple: electrons no longer tied to their atoms should, it would seem, be able to have all kinds of energy, but they haven't left the piece of metal. They no longer obey atomic laws but there are general rules for the metal as a whole that govern the behavior not of some one electron but of the whole electronic ensemble.

Stark Effect
The electron clouds alter their configuration in the presence of electric fields. Johannes Stark found that when a strong electric field is applied to a substance, the lines in the emission spectrum are split. This splitting has nothing at all to do with the twin lines caused by the spin. Yet there is something in common, which was demonstrated by Quantum Mechanics. The splitting of spectral lines corresponds to a splitting in the energy levels of the atomic electrons by the action of the <u>substantial electric field</u> on the atom's electric fields.

Stark won the 1919 Nobel Prize for Physics for his discovery of an optical Doppler effect and the observation that spectral lines split in an electric field (called the Stark Effect).

When a molecule is formed, the energy levels corresponding to the constituent atoms disappear. They break up, intermix, and shift up and down the energy scale to produce so-called **molecular levels of energy**, which now correspond to the entire molecule. But what relates to a molecule is more clearly expressed for a crystal where there are so many atoms packed close together repeating this packing throughout the crystal. Actually, a crystal is simply a gigantic frozen molecule.

The joint electric field of all the atoms of this 'molecule' splits up the energy levels of each one of them into an enormous number of very closely lying sublevels. The discreteness and distinctness of the permitted energy levels of the outer electrons vanish almost completely. It would seem therefore that an electron in a crystal can have any energy it wishes but with one very essential exception. The blank bands indicate energies which electrons in a metal cannot have. To these energies there corresponds a zero wave function and, accordingly, a zero probability of an electron finding itself in such a state.

These blank white bands of energy were termed *forbidden zones*, or bands. And even in the cross-hatched, so-called permitted bands, an electron is

Johannes Stark (1874-1957), Germany.

not allowed to have just any energy. There are separate energy levels in these bands as well. But each band has so many of them that they simply merge into a continuous sequence.

On these molecular energy levels, electrons still obey the Pauli's Exclusion Principle. Only two electrons are allowed on each energy level of the permitted zone of a metal. There is plenty of places, and more than enough levels of energy. Under normal conditions, all the electrons of a metal can settle in the lowest permitted zone, on the ground floor. Under the lowest level of permitted band lies the individual atoms and not to the atoms of the metal as a whole. Between the ground level of the molecular energy band and the individual atomic levels, there is a single rung equal in height to the first forbidden band. If knocked hard enough, an electron can be boosted from the individual atomic levels to the ground of the molecular energy level. But it is not allowed to get stuck in the forbidden band due to lack of energy.

The energy level below the molecular ground level is named *"valence zone"*, or valence band. And all the allowed bands of energy are called by the generic name *"conduction bands"*. The valence band is inhabited by some of the outer electrons that determine the valence (though these electrons are not yet free), while the ground floor and the upper stories are inhabited by electrons that participate in the conduction of electricity.

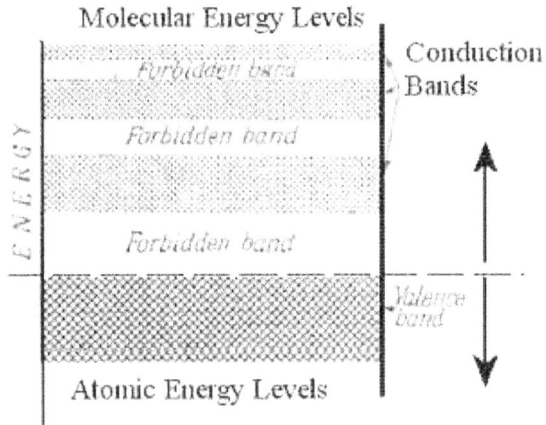

Insulators keep all their electrons in the valence band. Under ordinary conditions, their conduction band is empty, the first forbidden band is too broad for any of the electrons to find the energy needed to jump across it. But when the insulator is heated properly, the energy of oscillations of its ions at the lattice nodes becomes very great. This energy can be imparted to the electrons and these occasionally become energetic enough to jump up into the conduction band. The insulator then begins to conduct current. This is called heat *breakdown*.

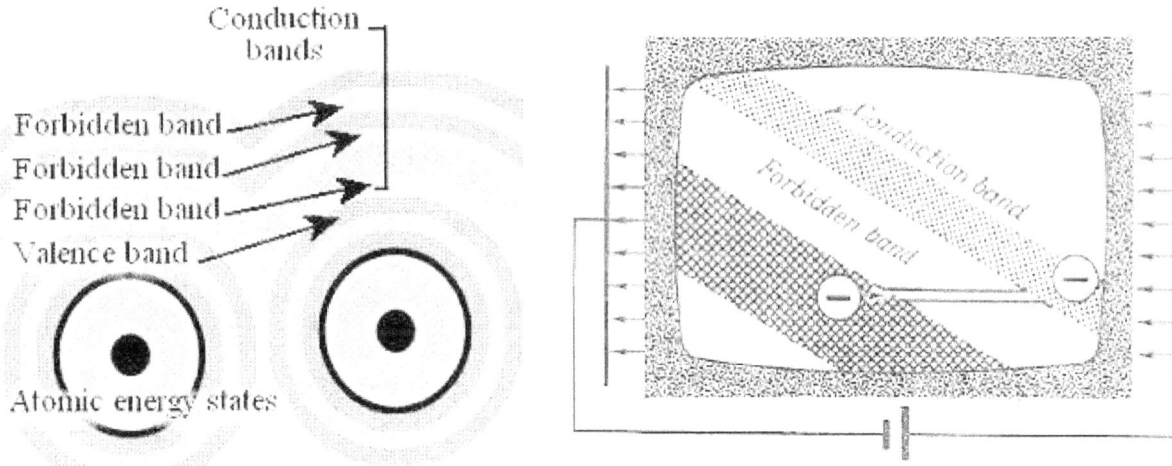

The breakdown causes the electron to leave its narrow atomic energy state and to enter the conduction band and become practically free. The energy required for its release is simply equal to the width of the forbidden band separating the valence band from the ground floor of the conduction zone. Also, an insulator becomes a conductor of electricity when a very strong electric field is applied to it.

Of course, the insulator isn't a metal, it's an ionic crystal. In metal, the electrons escape the metal altogether, whereas here, they only jump from the valence band to the conduction band. The electrons begin to seep through from the valence band across the barrier into the conduction band. First we get a small electric current: the probability of penetration is low and few electrons get into the conduction band. But this current, as it moves through the crystal, heats it. This heating, in turn, adds fresh flow of electrons to the conduction band and the current in the insulator builds up of its own accord. In just brief time, there is an electric rupture of the insulator.

This is accompanied by a simultaneous thermal breakdown- the insulator melts. But there is a more peaceable manner of generating electric currents in insulators. These currents are very weak and absolutely harmless. They are produced by illuminating ionic crystals. Photons strike the crystal, knocking electrons from the valence band into the conduction band. This is a real photoelectric effect, but there is no emission, it all takes place inside the insulator.

Conduction in Metal

Why does a metal filled with so many current carriers offer resistance to the flow of current?
A conductor is not a pipe; the walls are not rough to explain resistance. Electric currents have been known since Volta invented the electric batter in 1800. Ohm described the resistance law in 1826. Classical physics explained electrical resistance as thermal vibrations of ions that impede the motion of electrons. Obviously, the smaller the vibrations of ions, the lesser is the resistance. At the absolute zero of temperature, when the thermal vibrations of the ions cease altogether, electrical resistance should drop to zero. That is true in regards to *very pure metals* that are devoid of impurities.

The trouble lies with these impurities. As the temperature falls, the resistance of such 'dirty' metals does not tend to zero, but rather to some nonzero value which depends on the content and type of impurities in the metal. *The more impurities there are, the higher this residual resistance.* Classical physics doesn't distinguish an atom of the metal from an atom of impurity. At the same temperature they vibrate in the same manner and impede the electron motion in exactly the same way. In Quantum Mechanics, these different atoms in the lattice are distinguished very clearly almost as if they were of different colors.

How does Quantum Mechanics account for the electrical resistance?
De Broglie matter waves accounts for electron diffraction in the presence of obstacles with dimensions of the order of the wavelength of the electron wave. The electrons impinge on the outer layers of the atoms of the crystal, they then partially reflect and form diffraction rings. The electron current in a metal is no different from a beam of electrons. The electrons stream along in one general direction. The passage of electrons in a metal should be accompanied by an 'internal diffraction'. Diffraction has an interesting property. If there is the slightest deviation in the regularity of the objects scattering the waves, the clear-cut pattern vanishes and transmitted pattern is uniformly fogged. The scattering of the waves has become homogeneous. It is just such disorder that is introduced into the regular structure of a metallic crystal by ionic vibrations, and by the presence of impurity atoms. As a result, the waves of the electrons participating in the current are scattered in all directions.

As a rule, the impurity atoms have quite different dimensions and electron shells than the atoms of the metal. The impurity atoms distort the lattice. It is clear that these defects remain even when the atoms cease to tremble. Sure enough, the distortions introduced into a metallic lattice by impurity atoms are independent of the temperature and remain even at absolute zero. The scattering of electron waves on these lattice imperfections is the cause of the residual electric resistance of metals by virtue of de Broglie's matter waves.

Superconductivity
Thus, it turns out that metals are far from perfect as conductors of current. True, not all of them and not at all times and some are superconductors. A number of metals and alloys (as yet, just a few) begin to behave very strangely at extremely low temperatures. At just ten or so degrees above absolute zero, these substances suddenly lose practically all their electrical resistance. This phenomenon became known as **superconductivity**. Classical physics could not find an explanation for it. It is interesting to note that even the powerful Quantum Mechanics had to work hard for about thirty years before it came up with anything reasonable.

Bogolubov disentangling the mystery of superconductivity. At very low temperatures close to absolute zero, the interaction of the electron cloud with the ionic skeleton in a number of metals changes drastically due to certain peculiarities of structure. At the low temperature of superconductivity the electrons form into pairs. The electron pairs warded off the forces of individual ions with great ease as follows: The difficulties of the electron flow were reduced, and finally the electrical resistance of the metal fell off sharply. The wavelengths corresponding to electron motion in the metal are of an order of magnitude thousands and tens of thousands of times greater than the distances between ions. The wavelength of an electron pair is so much greater than the dimensions of the ionic obstacles in its path that the scattering of individual electrons, which accompanies the passage of current through a metal under ordinary conditions, disappears-and with it, the resistance to current. This ideal organization of the electron flow is maintained only so long as the temperature is sufficiently low. As the temperature rises above a certain limit, the clashes with ions break up the pairs into separate electrons. The balance of forces has changed and the electrical resistance of the metal is restored.

Semiconductors

A great number of things belong neither to conductors of electric current nor to insulators, but to semiconductors. Their semi- or intermediate properties have proved so valuable to the present technological revolution. The properties they possess are rather well known. Unlike insulators, semiconductors conduct current at room temperature, and unlike conductors, their electrical resistance does not increase with temperature, but falls off. The sharp dividing line

N. N. Bogolubov (Aug. 21, 1909- Feb. 13, 1992), Russia.

between insulators, semiconductors and conductors comprises of the gap that lies between them is the first forbidden band between the electron-filled valence band and the conduction band with its numerous unoccupied electron states.

In insulators, a great deal of energy is needed for an electron to climb out of the valence zone into the ground floor because the step is high. This energy may be obtained only at high temperatures or during thermal rupture. In semiconductors this step is much smaller. The energy the electrons require to make their way up into the ground floor is now obtainable at room temperatures. Thus, semiconductors begin to conduct current at ordinary temperatures. In other words, when even a weak electric field is applied to a semiconductor, a directed flow of electrons is set up in the conduction band.

In insulators, when an electron moves out of the valence zone into the ground floor it leaves behind a vacant room. The densely populated valence zone immediately begins a re-division of living quarters. Only one electron is permitted to move into the room; this is straightway done by one of the electrons close at hand. But it in turn leaves behind an empty room, which again, in turn, is occupied by a fresh electron. In jumping from room to room, these valence zone electrons imitate the freely moving electron on the ground floor. This traveling electron room was given the name of 'hole'.

The behavior of a "hole" is just the converse of that of the electron which left the hole-in an electric field. The hole moves in the opposite direction, like a positively charged particle. Another difference is that the hole moves in slower and larger jumps. At low temperatures all the electrons are securely trapped in the valence zone. As the temperature rises, however, more and more of them are released, the current increases and the resistance of the semiconductor diminishes. It is just the other way around for a metal conductor.

So far we have been talking about pure semiconductors. The current mechanism here is called *intrinsic conductivity*. Pure semiconductors, however, are of little interest to technologists. All the marvels that semiconductors are capable of come with what are called impurities produced in definite proportions. Accidental impurities are unwanted and all impurities are detrimental to electrical conductivity. Yet, these lattice imperfections are the key to the success of semiconductors as they alter the structure of the energy bands of a crystal in exceedingly sensitive manner that enables us to use crystals as optical and electrical *modulators*.

Every crystal has its own system of energy bands. However, impurity atoms do not alter the shape of the entire lattice but only in their immediate vicinity. The band pattern common to the whole crystal is appreciably modified in these areas. Additional allowed electron-energy levels appear in the forbidden band that separates the valence band from the conduction band. These levels originate from impurity atoms and are called local levels. While the amount of impurities in a metal also lowers its conductivity, in semiconductors, the electrical conductivity may be varied not only by the number of impurity atoms but also by the type of impurity atom, and the changes may be thousand-fold and million-fold.

The most common semiconductors are the chemical elements of Germanium (Z. 32) and Silicon (Z. 14). They are in the fourth group in the Periodic Table that is called *intermediate elements*. Germanium and Silicon are neither conductors nor insulators, they are typical semiconductors. The outermost shell of either atom contains four electrons. When the atoms are brought into a crystal, all these electrons form bonds with other atoms. At low temperatures, Silicon and Germanium do not conduct current. Impurities atoms such as Arsenic (Z. 33) (from the fifth group of the Periodic Table) could alter the conductivity Germanium to great extent. The Arsenic atoms will dislodge the Germanium atoms and occupy their places in the lattice. An atom of Arsenic has five electrons in the outer shell. Four of these are donated to form the chemical bonds of the displaced Germanium atom. The fifth electron in the Arsenic atom is left unemployed with its energy exactly corresponds to the local level in the forbidden band, but near the limit. Only a very little energy is needed to push this electron into the conduction band 10 to 15 times less than the height of the forbidden band itself. The arsenic atom is called a *donor* of the conduction electron. And the respective electron levels are called *donor levels*

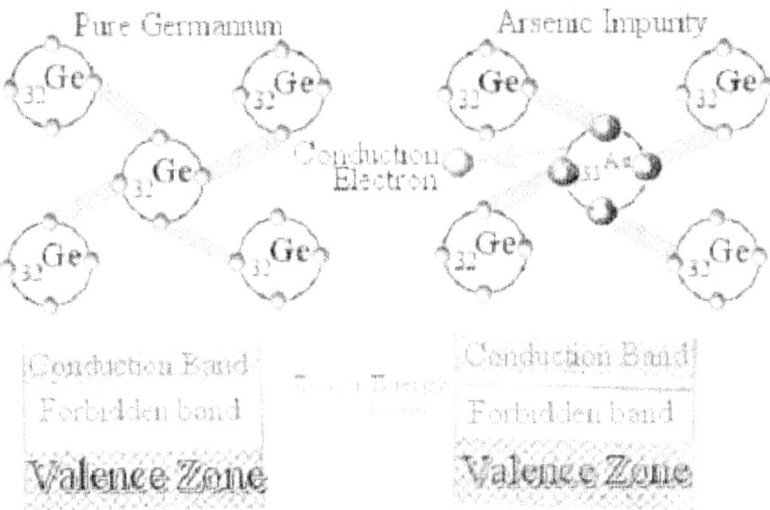

Born (Z. 5) is another interesting impurity. Boron is in the third group of the Periodic Table, which means that its outer shell has only three electrons. When Boron takes the place of a Germanium atom in the crystal lattice, it can handle only three of the four chemical bonds. The remaining fourth bond grabs an electron from a close lying neighbor, which does the same. This unoccupied electron room begins to move farther and farther away from the Germanium atom that first stole it from its neighbor

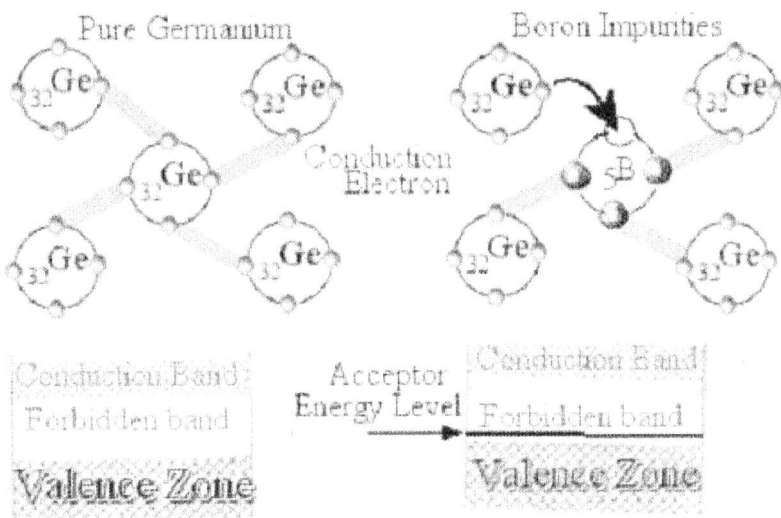

This process of migration of the empty room is called "migrating hole". Thus, the presence of Boron atom caused the electron to eject from the valence band, Thus Boron created local energy levels in the forbidden hand <u>near the valence band</u> where holes, not electrons, can occupy them. The Boron impurity atoms are called *acceptors*. The corresponding hole levels are known as *acceptor levels*.

Hence we have two types of electrical conductivity- by *electrons* or by *holes*-in accord with the type of atom that settles in the lattice of Germanium or Silicon. We have already seen that the electron levels in the conduction band are also separated one from the other, but that these distances between levels are so insignificant that the levels actually merge.

Mixing Boron and Arsenic atoms with Germanium results in two different types of conductivities depending on the ratios of As and B. Semiconductors with such double sets of impurities are capable of stopping the flow of current in one direction and of passing it in the opposite direction. Thus, they act as *rectifiers* and *modulators* or *amplifiers* replacing the big, cumbersome electronic valves. Photons striking a semiconductor knock electrons from the valence band into the conduction band. When illuminated, a semiconductor in a circuit begins to conduct current. Thus, semiconductors are capable of transforming light energy directly into electric energy and have thus replaced the cumbersome chemical photographic films. Silicon batteries in remote areas convert the solar energy into electricity. Semiconductor electric batteries have found applications in space exploration. It converts thermal energy directly into electricity.

CHAPTER 4:
THE INTERIOR OF THE NUCLEUS

Radioactivity

In 1895, the French chemist Henri Becquerel quite accidentally noticed that some substances are capable of clouding a photographic plate. Following up this discovery, Marie and Pierre Curie found that this property is possessed by the three chemical elements: Radium, Polonium and Uranium which lie at the end of the Periodic Table. This phenomenon was called *radioactivity*. Classical physics could not account for the mysterious radiation. There were three kinds of radiation: alpha, beta and gamma rays. Further studies demonstrated that alpha rays consist of positively charged particles. An alpha particle had a charge double that of the electron, while the mass was roughly four times that of the Hydrogen atom. Beta rays were undistinguishable from electrons. Gamma rays were extremely hard electromagnetic radiation, with a penetrating capacity many times that of X-rays.

In hindsight, the reader might imagine the challenges facing the early explorers of radioactivity. The X-ray was newly discovered in 1895. The mass spectrometer was recently invented by J. J. Thomson in 1897, who also discovered the electron in the same year. Then, in 1899, neither the proton nor the neutron was discovered and neither the radio broadcasting nor the automobile was in public use. Geographically however, the electron was discovered in England,

the X-ray in Germany, and the radioactivity in France. Then, the fastest mean of communication was the first telephone line between Paris and London, laid in 1891 by Anglo-French Telephone Cable. The beginning of the nineteenth century was truly the dawn of modern physics.

In 1911, the English physicist Rutherford working together with his pupil Bohr expounded the planetary model of the atom in which electrons, like planets, revolve about the atomic nucleus. It gradually became clear that the source of radioactivity was the nucleus. With respect to the alpha particles, this was obvious from the very start. There is no place in the atom for them except the nucleus, which contains practically the entire mass of the atom. On the other hand, electrons exist in the outer atomic shells. We also frequently find photons (quanta of electromagnetic energy) flying out of these shells. But, **when an atom emits beta rays, it does not become ionized, it does not acquire an electric charge and thus its electronic structure remains unchanged**. Further calculations of energy corresponding to photons of visible light and X-rays associated with jumps in electron shells demonstrated that this energy is only a fraction of the photon energy of gamma rays. This strengthened **the idea that the beta and gamma radiation originates in the atomic nucleus.**

The Nucleus Of The Atom

In 1919, Rutherford bombarded various gases with alpha particles and found that every once in a while an alpha particle would strike the nucleus of an atom and disarrange it. The alpha particle could knock protons out of Nitrogen nuclei and merge with what was left behind. It follows then that Rutherford had successfully carried through the first man-made nuclear reaction:

Nitrogen-14 + Helium-4 (alpha particles) --> Oxygen-17 + Hydrogen-1

This is a true example of **transmutation**, the conversion of one element to another. In a way, it was the climax of the old alchemical goals but it involved elements and techniques of which the alchemists had never dreamed. Over the next five years, Rutherford carried through a number of other nuclear reactions involving alpha particles. What he could do was limited because radioactive elements provided alpha particles of only moderate energies. To accomplish more, much more energetic particles were required.

Acceleration of charged particles to greater energies had just started. The English physicist John Douglas Cockcroft (1897-1967) and his co-worker, the Irish physicist Ernest Thomas Sinton Walton (1903-1995), were the first to design an accelerator capable of producing particles energetic enough to carry through a nuclear reaction, accomplishing this in 1929. Three years later, they bombarded Lithium atoms with accelerated protons and produced alpha particles through the following nuclear reactions.

Hydrogen-1 (protons) + Lithium-7 --> 2 Helium-4 (alpha particles)

That same year, Rutherford observed the first nuclear transformation and found that the nuclei of atoms of a single chemical element can have different masses. Calculations showed that these masses differed by an amount that was a multiple of, or very close to, the mass of an atom of Hydrogen. Such nuclei became known as *isotopes*.

Incidentally, one nuclear particle, the proton, was definitely established. We could now surmise that the nucleus consisted of those particles that appear in radioactive disintegration: alpha particles and electrons. Alpha particles seemed to have the same properties as Helium nuclei. Yet there were still lighter nuclei, those of Hydrogen.

Thus, the Hydrogen nucleus should be the smallest building stone in the nuclear structure of elements. Since this was the most common elementary particle, it got the Greek name proton. Now we can begin building our model of the nucleus. We must take into account the basic rule-**the charge of the nucleus must be equal to the collective charge of all the electrons in the outer structure, but with opposite sign (positive).** Otherwise the atom would not be neutral the way it is.

We also know the nuclear masses: they are roughly equal to the masses of the corresponding atoms minus the masses of their electronic shells. So we have our starting hypothesis: nuclei consist of protons and electrons. The nucleus of Hydrogen has one proton, and there are no electrons at all. The Helium nucleus has four protons and two electrons; its charge is thus +4-2 =+2, its mass is just a little bit more than four times the Hydrogen nuclear mass. We know that the electron is almost 'weightless' compared to the proton-nearly 2,000 times lighter.

The Lithium nucleus with mass 7 and charge +3 consists of 7 protons and 4 electrons, the Boron nucleus with mass 11 and charge +5 consists of 11 protons and 6 electrons, Nitrogen (14 and +7, respectively) consists of 14 protons and 7 electrons; Oxygen (16 and +8) consists of 16 protons and 8 electrons, and so forth.

That trend held true for the light nuclei. But as soon as we move into the region of medium and large structures, agreement breaks down. Iron with nuclear mass 56 and charge +26 requires 56 protons and 30 electrons; for the Uranium nucleus with mass number 238 and charge +92, we must have 238 protons and 146 electrons. Thus, the regularity in nuclear construction breaks down. Then there is difficulty in figuring out where isotopes come from. Trouble with nuclear spins too sets in from the very start. Their **total spin must be equal to the sum of the spins of the constituent particles**.

For example, for the nucleus of heavy Hydrogen (Deuterium) which is generated via this scheme by two protons and one electron, the total spin should at least equal to three proton spins (proton and electron spins are equal). Actually, it is equal to two proton spins. And this is not the only discrepancy. There must be something wrong in our method of building nuclei. The nuclear electrons have the job of building up the charge of the nucleus so that it corresponds to the experimentally observed value. But they have a still more important function. Protons repulse each other being of the same charge, just like electrons in the shells of the atom. The electrons are needed to hold the protons together. A simple calculation shows that **the nucleus will require much more electronic glue than we have in our method of construction**. There are still other, more convincing, objections to the presence of electrons in nuclei. Thus, there are serious doubts about nuclei consisting of protons and electrons.

The Neutron

He worked under Professor Rutherford and under Professor H. Geiger. He joined back up with Rutherford in Cambridge and accomplished the transmutation of other light elements by bombardment with alpha particles and in making studies of the properties and structure of atomic nuclei. In 1932, Chadwick discovered the neutron. It was obvious that the nucleus of the atom was a composite that gave radioactivity-particles flying out of the nucleus and the nucleus continuing to exist.

Then in 1932, the neutron made its appearance. That was followed by Werner Heisenberg and the Soviet physicists D. Ivanenko and I. Tamm who advanced the hypothesis that nuclei are made up exclusively of protons and neutrons. In the light elements (approximately up to Calcium, Z. 20), the number of protons and neutrons in the nucleus was about the same. After that, the number of neutrons grew faster than the protons, and this continued -the farther away, the greater the difference. In the Uranium nucleus, with mass number 238 and 92 protons, we find 146 neutrons. **Variation in the neutron architecture of the nucleus led to isotopes**-variations of a single element. The nucleus of Tin, for instance, has ten stable isotopes.

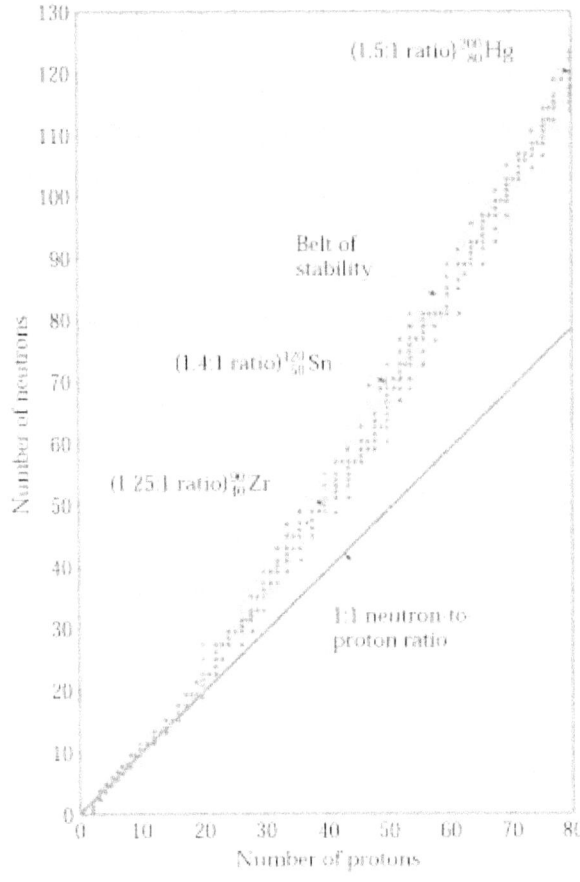

James Chadwick (October 20, 1891-July 24, 1974), England.

It was readily seen that the Heisenberg- Ivanenko-Tamm hypothesis is in excellent agreement with the data on nuclear masses and charges. According to this theory, the Hydrogen nucleus consists of a single proton, the Helium nucleus of mass number 4 (Helium-4) consists of 2 protons and 2 neutrons, the Lithium-7 nucleus has 3 protons and 4 neutrons, the Boron-11 nucleus has 5 protons and 6 neutrons, the Nitrogen-14 nucleus is composed of 7 protons and 7 neutrons, the Oxygen-16 nucleus is made up of 8 protons and 8 neutrons, and so on and on. And so on and on, right to the end of the Table this time.

The newly discovered neutron has a mass almost exactly equal to that of the proton and is electrically neutral-it has no electrical charge. By what right does it occupy the place of the electron in the nucleus? The electron at least could hold the protons together. How can an uncharged neutron do this? At this point we find out that the electrical forces of attraction are not enough to account for the stability of nuclei. Nuclei are extremely stable against chemical means, enormous pressures or temperatures or the strongest electric fields.

Nuclear Forces

The neutron plays the role of cement in holding together the protons of the nucleus. But by what force, we ask. It can't be electrical, for the neutron is neutral. Two years after the discovery of the neutron, I. Tamm and the Japanese physicist Yuka-wa put forward a brilliant idea that there are very strong specific nuclear forces, exchange forces of attraction, of very short range operating between protons and neutrons.

Those are the forces that hold two Hydrogen atoms together or the atoms of Nitrogen, Oxygen and many others in rather stable molecules. In these molecules, the atoms are continually exchanging their electrons, and this holds the atoms together. The proton and the neutron are two different particles. The nucleus hasn't got any electrons. There must exist some mysterious reaction between proton and neutron.

In 1934, a calculation made by I. Tamm showed that electron exchange yields too small a cohesive force for nuclear particles. I. Tamm postulated that the proton and neutron are actually not very different and have much in common, that they can convert into one another: the proton into the neutron and the neutron into the proton. Two years before, in 1932, the transformation of an electron and positron into gamma-ray photons had been established. But this phenomenon was of an utterly different nature.

Reasoning further, physicists figured that if two particles can convert into one another they should exchange something in the process. The proton acquires this 'something' and turns into a neutron; and when a neutron loses it, a proton appears. Then again, there could be a different kind of exchange in which the neutron acquires something and the proton does the losing.

Starting from the fact that nuclei are extremely stable structures and also that the exchange forces between protons and neutrons would have to operate over the extremely small distances between the particles, **Yukawa postulated the existence of a new exchange particle with a positive or a negative charge equal to the magnitude of the proton charge (or electron charge) and a mass approximately 200 to 300 times greater than the electron mass.** The proton and the neutron are roughly 1800 times more massive than the electron. So the mysterious particle would have a mass somewhere in between the two. Hence the name: **meson** from the Greek meaning `medium'.

Then we get the following picture of nuclear exchange. A proton emits a positive meson, loses its positive electric charge and converts into a neutron. A neutron picks up a meson and turns into a proton. But a neutron may emit a negative meson and become a proton in a different way. And this meson, when captured by the proton will convert it into a neutron in still another way.

Igor Tamm (8 July 1895-12 April 1971), Russia

Hideki Yukawa (23 January 1907-8 September 1981), Japan.

Discovery Of The Meson

Experiments with radioactive nuclei did not yield meson. Mesons never leave nuclei. As if mesons preferred to carry on their modest yet important work and never show up. Then physicists turned to that great source of information about nuclear particles, the **cosmic rays**. Within the year the meson was discovered. It has a mass of roughly 200 electron masses: One of the most remarkable attainments in physics ever. However, the discovered mesons were extremely indifferent to neutrons, and only bowed to protons within the ordinary framework of electrical interaction.

Powell, in 1947, was able to discover another brand of meson. This particle was again a meson, but a different one, not 207 electron masses but 273. There was no mistake this time. The new meson (called a **π-meson** or **pion** to distinguish it from the indifferent **μ-meson** or **muon**) interacted strongly with nuclear particles. When it had considerable energy in flight, it could even break up nuclei that it encountered. Thus, **nuclear forces are now attributed a meson exchange between protons and neutrons**

Nuclear Stability And Abundance

The newly discovered nuclear forces have the extremely short range of action. The exchange forces in, molecules begin to operate at interatomic distances of the order of the dimensions of the atoms-hundred millionths of a centimeter (10^{-8} cm). Nuclear exchange forces have a range tens of thousands of times shorter (10^{-12} cm). They begin to operate only at distances close to the dimensions of the nuclear particles themselves. And so, quite reasonably, they can exist only inside the nucleus and never display action outside. Nuclear forces are the strongest yet discovered. Not only do they completely suppress the mutual antipathy of the protons, which is very great at such small distances, but even hold them tight in a

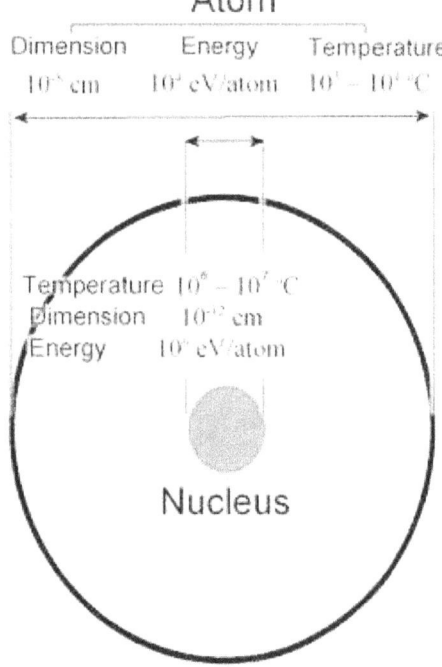

Cecil Frank Powell (5 December 1903- 9 August 1969), Italy.

stable structure.

The nuclear strength or stability is the binding energy that must be imparted to an assembly of particles in order to break it up into its constituents. Naturally, the more particles there are in a system, the greater this energy must be. To describe stability, we usually take the binding energy per particle. This energy is measured in special units called *electron-volts*. One electronvolt is the energy acquired by an electron when passing through a potential difference of one volt in an electric field. The bonds between molecules of many substances are of the order of hundredths of an electron-volt per molecule. To break down these molecules into individual atoms, a much bigger energy is needed, roughly ten electron-volts per atom (≈ 10 eV/atom). This corresponds to impressive temperatures ranging up to thousands and tens of thousands of degrees ($\approx 10^{3} - 10^{4}$ °C)

91

To decompose atoms into their constituent electrons and nuclei is still more difficult. We know that atomic electrons have different energies corresponding to their coupling with the nuclei. This energy range is from tens to thousands of electron-volts ($\approx 10^4$ eV/atom). Nuclear particles have binding energies in the millions of electron-volts ($\approx 10^6$ eV/atom) Now it is clear why nuclei are unaffected by even the very strongest of non-nuclear forces. Even if two nuclei collide with speeds of thermal motion at thousand degree temperatures, the effect will be hardly more than that of a rubber ball bouncing against a wall of granite.

Studying the nuclear structure of elements led to the determination of the binding energy and the stability of the various nuclei. This nuclear binding energy is plotted against the mass numbers of the nuclei as follows.

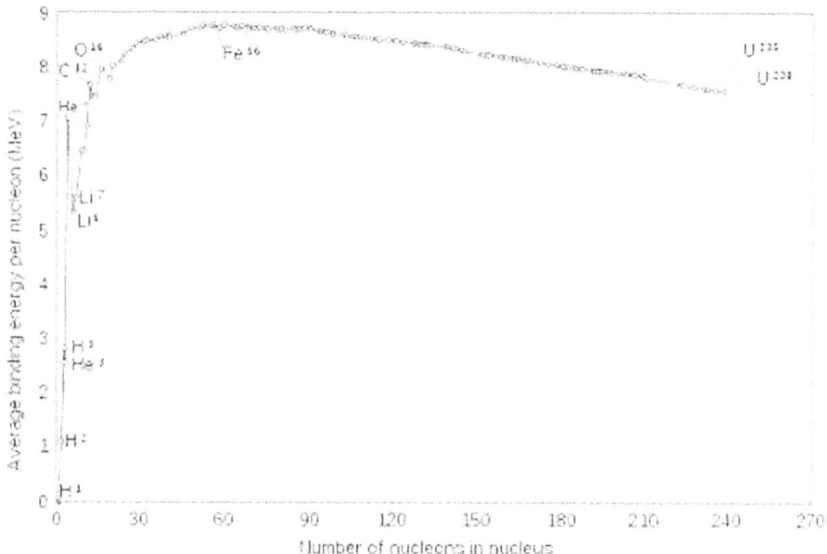

The above graph yields the semi empirical formula for nuclear binding energy per nucleon for a nucleus with a number A nucleons, including Z protons and N neutrons as follows.

$$B.E./A = a - b/A^{1/3} - cZ^2/A^{4/3} - d(N - Z)^2/A^2 \pm e/A^{7/4}$$

where the binding energy is in MeV for the following numerical values of the constants:
$a = 14.0$; $b = 13.0$; $c = 0.585$; $d = 19.3$; $e = 33$.

The first term a is called the *saturation contribution* and ensures that the **B. E.** per nucleon is the same for all nuclei to a first approximation.

The term $-b/A^{1/3}$ is a *surface tension effect* and is proportional to the number of nucleons that are situated on the nuclear surface; it is largest for light nuclei.

The term $-cZ^2/A^{4/3}$ is the *Coulomb electrostatic repulsion*; this becomes more important as Z increases.

The *symmetry correction* term $-d(N-Z)^2/A^2$ takes into account the fact that in the absence of other effects the most stable arrangement has equal numbers of protons and neutrons; this is because the *n-p* interaction in a nucleus is stronger than either the *n-n* or *p-p* interaction.

The *pairing term* $\pm e/A^{7/4}$ is purely empirical; it is "+" for even-even nuclei and "-" for odd-odd nuclei.

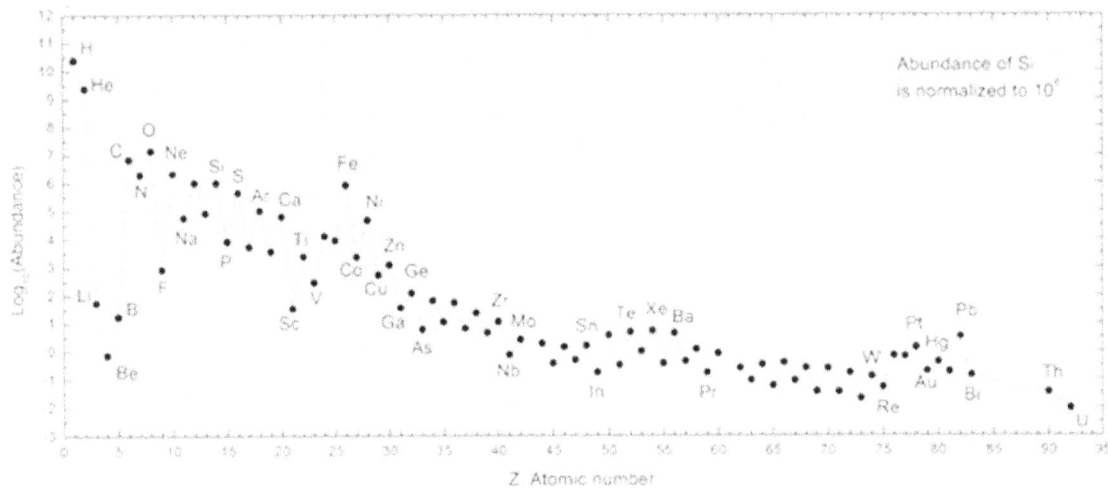

Abundances of the chemical elements: Solar system abundances. Hydrogen and Helium are most common. The next three elements (Li, Be, B) are rare because they are poorly synthesized in stars. The two general trends in the remaining stellar-produced elements are: (1) an alternation of abundance in elements as they have even or odd atomic numbers, and (2) a general decrease in abundance, as elements become heavier.

Comparing the distribution of the binding energy with the abundance of elements, we notice that the highest peaks of the abundance curve correspond to nuclei of Helium-4, Carbon-12, Oxygen-16, and a number of others. All these numbers are multiples of four, as if these nuclei consisted not of protons and neutrons separately, but of alpha particles straight off.

Most noticeable breaks in the binding energy curve correspond to peaks in the abundance curve. The more stable the nuclei, the greater, generally speaking, are its abundances in nature. The world of atomic nuclei nature has set up its own law of natural selection. The most abundant are those whose neutron and proton numbers are 2, 8, 20, etc. Thus, groups of two protons and two neutrons are indeed extremely stable even in the world of atomic nuclei. The nuclear forces operating between these numbers of particles become saturated.

The Helium nucleus, for example, refuses to accept anyone else into the family. Sure enough, this nucleus is the most inhospitable one of all. There aren't even any nuclei -with mass number 5 (two protons and three neutrons or three protons and two neutrons). If we exclude the Hydrogen nucleus (which consists of only one proton and therefore has no nuclear forces operating in it at all), the **Helium nucleus is the most stable nucleus in nature**.

Nuclear forces act independently from electric charges. They are attraction forces between a proton and a neutron as between a pair of protons or a pair of neutrons. The exchange forces that form these strong nuclear structures bind the protons and neutrons together. There must be a limit to nuclear forces since particles retain their individual entities and do not merge into one. Thus, powerful forces of nuclear repulsion were proposed that do not permit the particles to penetrate into one another. This is the so-called lower limit to the range of action of nuclear forces.

The upper limit of nuclear binding forces is of the order of the dimensions of the nuclear particles themselves. This explains the general trend of the binding-energy curve, which falls with the increase in mass number of the nuclei. In a light nucleus that has few protons and neutrons, each particle may be linked to all the others by nuclear forces. The nuclear forces in light elements act directly with its immediate neighbors. Such nuclei begin to lose their stability, all the more so since there is a general increase in the repulsive forces of the protons, which act in opposition to the nuclear forces, as the number of protons increases. The largest and heaviest nuclei at the end of the Periodic Table of elements are rather unstable and hence eject radioactive radiations. That is due to the saturation of the exchange forces that form these strong nuclear structures.

The great number of radioactive nuclei both at the beginning and in the middle of the Periodic Table is man-made. Bombarding originally stable nuclei with neutrons disturb their binding energy by overloading them with extra particles. The irradiated nuclei return to different final states than the original state by various routes of radioactive decay. A

93

nucleus is agitated by the addition or removal of extra nucleon and it responds by ejecting electrons and gamma photons until it gets transformed into an entirely different nucleus. Underlying this phenomenon, called *artificial radioactivity*, is the tendency of all nuclei towards stability.

Alpha Particle Decay

The alpha radioactivity, or the alpha decay of nuclei, which was discovered even before the neutron was, raised few problems: (1) most natural radioactive elements eject alpha particles while protons and neutrons do not eject separately. The combined atom of the alpha particle must have greater ease of seeping out of the nuclear well. (2) Alpha particles could not accept any extra nucleon despite the fact that elements heavier than alpha particles do exist. There must by a reason for alpha particles that make them self content and sufficient.

To explain the privileged status of alpha particle, the exchange of **π-meson** or **pion** was postulated. A neutron ejects a negatively charged π-meson and converts into a proton in the act, while the latter absorbs this meson and in a minute fraction of time turns into a neutron. In the case of a tetrad, there are two protons and two neutrons. When a meson is ejected by a neutron and some tetrad is captured by a proton of an adjacent tetrad, there will be two violations of the conservation laws. The first tetrad will have three protons and one neutron, and the neighboring one will have three neutrons and one proton. But, according to the Pauli principle, protons and neutrons have the same spin as an electron and so have to abide by all the restrictions imposed on electrons. The Pauli principle forbids more than one particle having a given sense of spin in a given state. The alpha particle is extremely stable because the two protons and two neutrons in it, each occupy a single energy level-the lowest possible. The two protons are on one level, and the two neutrons are on the same level. This is possible due to the fact that at each instant of time, a proton and neutron in the nucleus appear as actually different particles.

Now, if a tetrad has three protons, then one of them will simply have to violate the stringent Pauli's Exclusion Principle or will have to occupy a state of higher energy, which is to say a lower binding energy, i.e., an unstable state. The unstable tetrads immediately release the meson and return to two ordinary tetrads. But this instantaneous exchange between the tetrads establishes a mutual bond between them. The tetrads become less isolated from each other. The farther we go away from the light nuclei, the weaker are these effects of tetrads on stability.

Yet the heavy nuclei again exhibit a definite influence of tetrads. Particles on the periphery of such nuclei can interact only with their immediate neighbors because the nucleus has become so large. Apparently, near the surface of nuclei we find particles forming into tetrads again since this are the most stable configuration. This might explains the ejection of only tetrads (alpha particles) from heavy nuclei, and not protons or neutrons. Since the nucleus constitutes a potential well of height equal to the binding energy and width determined by the range of action of the nuclear forces, the ejection of an alpha particle from a radioactive nucleus is similar to the tunnel effect through the forbidden band of the atomic potential well.

In light nuclei the height of the barrier for ejection of alpha particles is very great (a large binding energy), while in heavy nuclei the barrier is low (a considerably smaller binding energy). On the other hand, the barrier height for ejection of alpha particles in heavy nuclei is much lower than that for the 'individual' release of protons and neutrons.

Nuclear Energy Levels

The nucleus is very much like the outer structure of the atom, where we found filled and closed and stable electron shells and the formation of noble elements.

After the discovery of the proton-neutron structure of the nucleus, the nucleus was considered as more or less evenly clouded out nuclear matter filling this tiny volume of space in the form of a cloud of protons and neutrons. However, the discovery of the saturation of nuclear forces and of the phenomenon of alpha disintegration seemed to indicate that the nuclear matter is not quite formless. The particle tetrad occupies the lowest energy position in the nucleus and that it is the most stable of all the nuclear blocks. To this position there corresponds a single general level of energy at which we find two protons and two neutrons with spins in opposite directions. The second tetrad of particles occupies a different energy level, the third, a third level, etc. As the number of particle tetrads increases, higher and higher energy levels in the nucleus are filled, very much like the electrons in atoms. But not all nuclei consist of tetrads. And this means that in nuclei with numbers of particles that are not multiples of four, the corresponding energy levels will not be fully, occupied.

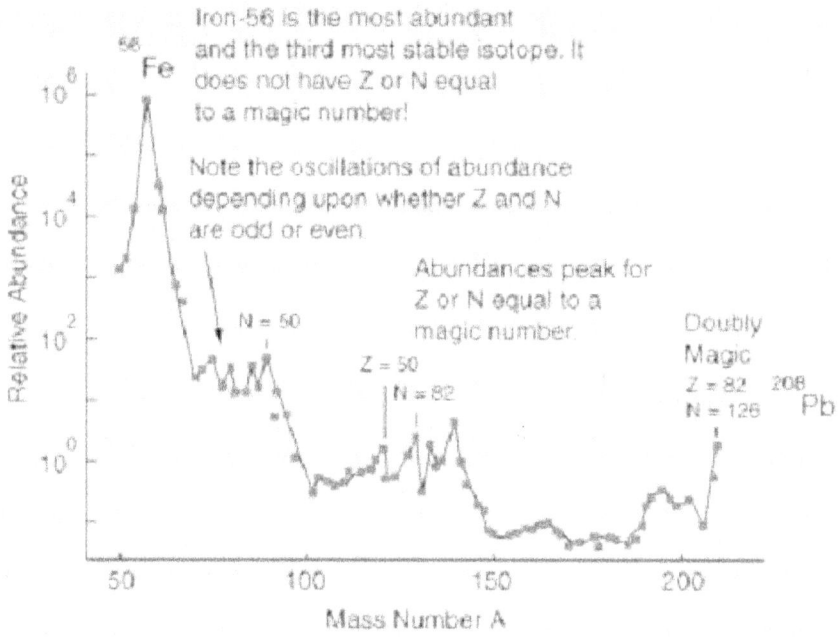

Iron-56 is the most abundant and the third most stable isotope. It does not have Z or N equal to a magic number!

Note the oscillations of abundance depending upon whether Z and N are odd or even.

Abundances peak for Z or N equal to a magic number.

N = 50

Z = 50

N = 82

Doubly Magic
Z = 82   208
N = 126   Pb

Relative Abundance

Mass Number A

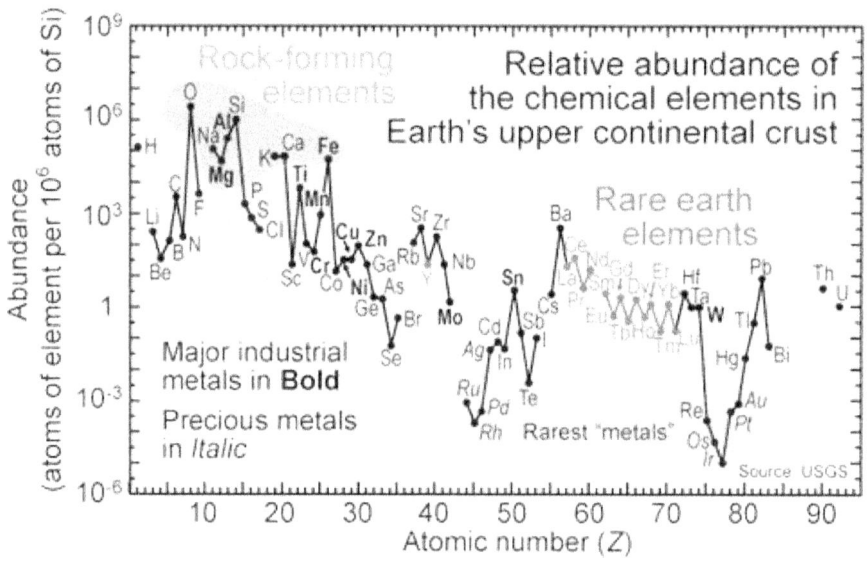

Relative abundance of the chemical elements in Earth's upper continental crust

Rock-forming elements

Rare earth elements

Major industrial metals in **Bold**

Precious metals in *Italic*

Rarest "metals"

Source USGS

Abundance (atoms of element per 10^6 atoms of Si)

Atomic number (Z)

In order to examine the shell structure of the nucleus, we start with the curves of stability and abundance of nuclei, take a few of the very highest peaks, and calculate the numbers of protons and neutrons in the nuclei that correspond to each. The first is Helium-4; its nucleus is an alpha particle and consists of 2 protons and 2 neutrons. Then, comes Oxygen-16 with 8 protons and 8 neutrons. This is followed by Calcium-40 with 20 protons and 20 neutrons, and so on. Finally, at the right-hand end of the curve is the last high peak that belongs to lead-208. Here, the nucleus has 82 protons and 126 neutrons. (To these we must add the tin nucleus with 50 protons, which is so stable that nature devised ten stable isotopes, whereas other proton numbers have only 2 to 5 stable isotopes.)

So, we now have the most stable nuclei with proton-neutron numbers 2, 8, 20, 50, 82, and 126. 200. Note that these nuclei are sort of counterparts of the atoms of the inert elements with 2, 10, 18, 36, 54, and 86 electrons. Both-each in its own world-are record-holders of stability. These proton and neutron numbers were named *'magic numbers'*, for their shared features between nuclei and electronic shells of atoms. The two worlds that live by entirely different laws, yet exhibit this common structural feature. However, a comparison of the magic numbers with the electron numbers in the most stable atoms shows up a definite difference. These numbers coincide only for Helium, which holds all records of stability in both worlds. It is no accident that these numbers diverge since the governing forces in the nucleus and in the outer electronic cloud are different as well. Still, **there is something like a system of shells in the nucleus**.

There is also experimental corroboration. Let us take a look at the Potassium atom (Z. 19). It is univalent, which means that it has one electron outside the filled and closed shell of the inert argon atom. The total spin of the electronic structure of the Potassium atom is equal to the spin of this valence electron. The spins of all the other electrons are in pairs and in opposite directions so that they cancel and the sum is zero. Comparing the spin pf Potassium-19 with nucleus of the Oxygen-17 isotope (which has one neutron over and above four tetrads of particles), we find that the spin of the nucleus of Oxygen is exactly equal to the spin of this extra neutron. Further, the experimentally measured spins of nuclei are in excellent agreement with those predicted on the basis of the model of nuclear shells. Therefore, the nucleus must have energy levels or shells.

Gamma Rays Decay

Like x-rays in the case of atomic electron shells, gamma-rays offer greater insight into the nuclear shell structure. The spectra of nuclear gamma rays were found to consist of separate lines. Thus, nuclear particles can strictly have only very definite energies. Thus, nuclear particles have to exist in specific energy states. The transition of particles between such states would then give rise to gamma rays. These levels are predicted by the Schrödinger equation for all connected assemblies of particles, which naturally include the atomic nucleus as well. In the case of an atom, the interaction of particles is described by Coulomb's law for the mutual repulsion of electrons and their attraction to the nucleus. But the law of nuclear forces is still unknown. The only access to the governing nuclear laws is the separation and intensity of the spectral lines of the gamma rays. To emit a gamma photon, the nucleus has to pass from a stable state with least possible energy to a state with more energy, which by analogy with the atom is called an excited state. The gamma photon is emitted when the nucleus returns to its original state or to some other stable state.

Nuclear forces are millions of times stronger than electrical forces. For this reason, the distances between energy levels in the nucleus are usually much greater than the energy distances in the electronic structure. It is natural to expect that the gamma photons too have to be just as many times as energetic as the photons of light. Gamma rays have the shortest wavelength of all known radiations.

The nucleus can also give up its surplus energy directly to the electron cloud. But this energy is so great that some of the electrons are fired out with very considerable velocities. This phenomenon competes very successfully with the direct emission of gamma rays and is called *internal conversion*.

As far as we know, the nucleus has no core that could be surrounded by nuclear particles. The closed groups in the nucleus consist of quite different numbers of particles than in the outer part of the atom. Finally, nuclear shells would have to be of two kinds, proton and neutron.

There is some justification in speaking of shells as concerns only the light nuclei which consist of a few nuclear particles. But as the nuclei increase in size, the separate energy states lose their individuality and the nucleus ceases to obey quantum laws, loses all features of similarity with the atom.

Nuclear Fission

Shortly before the outbreak of World War II, the nucleus was pictured as an outwardly homogeneous mass without any ordered structures, has mobile and fluid boundaries, kept intact due to a certain surface tension of the nuclear liquid on the boundary of the drop. The nuclear particles are bonded by forces of attraction not countered by any other forces outside the liquid drop. The nuclear forces hold the nuclear liquid in this drop. The particles of a nucleus are packed in thousands of millions of times more tightly than are the molecules of an ordinary liquid. A nuclear droplet the size of a drop of water dripping from the tap would weigh a good ten million tons.

Yet we know that the properties of bodies are very greatly dependent on their densities. Change the density of a gas one thousand times and it becomes a crystal obeying utterly different laws. It should be clear, then, that we cannot speak of any kind of internal similarity between ordinary liquids and the nuclear liquid. There is too great a difference in their density-thousands of millions of times and the forces acting between nuclear particles differ radically from those acting between molecules.

In 1939, the fission or break-up of Uranium nuclei was discovered. Independently, Niels Bohr and the Soviet physicist Ya. Frenkel advanced the liquid-drop model of the nucleus to account for its fission. The neutron bombardment of the nuclear liquid is immediately distributed among all the nuclear particles. A new nucleus is thus formed. Bohr called it a compound nucleus. But this state doesn't continue for long. One of the particles finally gets a strong enough bump to knock it across the potential barrier at the boundary of the nucleus, and leaves it. If the emerging particle is different from the one that entered, the whole sequence of events is called a *nuclear reaction*. The name is justified in that the initial nucleus differs from the terminal nucleus. Just like in chemistry where the initial substances differ from those produced in the chemical reaction. That resembles random thermal motion of molecules in a liquid drop. From time to time, separate molecules evaporate from the drop. Nuclear radiation is thus much like the 'evaporation' of particles from a nucleus heated up by the impact of an outside particle.

The surface tension in a nuclear liquid drop is due to the nuclear forces of attraction. The larger and more massive the nucleus, the weaker are these forces and the more feeble is their hold on the nuclear particles. And in heavy nuclei, even relatively weak jolts can build up dangerous oscillations on the surface. Neutron colliding with the massive and rather unstable Uranium nucleus (recall that due to their instability these nuclei are radioactive) causes such weakening of the nuclear surface tension. The slightest jolt will break up a Uranium-235 nucleus, say in a collision with a thermal neutron, which is a neutron with energy hundreds of millions of times less than that typical of atomic nuclei.

In 1930-1931, Frenkel showed that neutral excitation of a crystal by light is possible, with an electron remaining bound to a hole created at a lattice site identified as a quasi-particle, the exciton. Bohr and Frenkel presumed that nuclear fission is due to a similar deformation of the nuclear surface when neutrons impinge on heavy unstable nuclei.

It well understood that neutrons could penetrate the Coulomb's barrier with impunity and thus settle inside the nucleus of the atom. But, why do massive nuclei prefer to fall into large pieces and not evaporate out individual particles, as in the case of artificial radioactivity in nuclei of small and medium mass?

We assume that the potential barrier which separates the nucleus from the outer world has two sides. On the inner side, the nuclear barrier is more sloping for protons than for neutrons. For protons, the nuclear fence consists of the nuclear forces minus their mutual Coulomb's repulsion. Thus, protons could seep easier of the nucleus compared to neutrons. For neutrons there is no outside barrier because they are electrically neutral. On the contrary, there is a well into which they can fall-when they fall into a nucleus, they usually stay there. Therefore, if a proton wants to get [into] a nucleus, especially into a heavy, multiproton nucleus, it has to have enormous energy ranging up to hundreds of millions of electron-volts. A neutron doesn't need any energy at all. Neutrons of very low energy (even thermal neutrons with energies of hundredths of an electron-volt) can enter a nucleus.

Yakov Il'ich Frenkel (10 February 1894- 23 January 1952), Russia.

The nucleus of Uranium-235 could, without any danger to its stability, accommodate three more neutrons to form a nucleus of Uranium- 238. The new neutron neither overloads the nucleus nor adds appreciable energy. **The nucleus of**

Uranium-235 is fissionable by a neutron of only a very definite energy. The energy limits correspond to the distance between the energy levels related to the stable and close lying excited states of the Uranium-235 nucleus. Neutrons whose energies correspond to the energy difference of the two states just mentioned are most effective in exciting Uranium nuclei.

In the Uranium-235 nucleus the energy distance between excited and stable states is very small. Once in an excited state, this nucleus should act like the light nuclei and emit a gamma photon and some particle, and then return to the same or some other stable state. But that is not what happens. The heavy nuclei eject alpha particles (tetrads) instead of separate particles. This is due to the fact that the potential barrier for the emission of alpha- particles is considerably lower than for the ejection of individual nuclear particles. The barrier for such large 'blocks' as nuclear fragments in the fission process is very low in the case of the Uranium- 235 nucleus.

Once in the excited state, this nucleus is able to wobble over the tiny fission barrier and break up into fragments. A very similar situation is found in the case of molecules. The energy required to eject even one single electron from the molecule is rather substantial. But the energy needed to split the molecule into separate atoms is much less. In chemical reactions, molecules do not break up into electrons but into atoms or into groups of atoms (radicals).

The fission of Uranium-238 nuclei by neutrons is very much like that of Uranium-235. But in Uranium-238, the excited state is separated from the stable initial state by a rather broad energy range of a good million electron-volts. Thus, very fast and energetic neutrons are needed to raise such nuclei to the excited level.

The upper limit on the number of nucleons in atomic nuclei is well demonstrated by the limited numbers of elements in nature. The heavier a nucleus, the less stable it is. But even a Uranium nucleus exists thousands of millions of years on the average before spontaneously getting rid of its 'extra' alpha particles and reducing to a more stable state. Heavier nuclei than Uranium can live for quite some time before ejecting an alpha particle. Heavy nuclei have very low nuclear barriers that lead to a perceptible probability of spontaneous splitting without any excitation. The spontaneous fission of heavy nuclei was discovered by the Soviet physicists Flerov and Petrzhak. The heavier the nucleus and the more particles there are, the greater the probability of such fission.

This is very rare in Uranium nuclei, the probability is practically zero. But for Californium (Z. 98), the mean lifetime of nuclei for spontaneous fission is just a few years. And finally we have a nucleus where the barrier to fission simply vanishes. A nucleus of this kind should have no resistance to fission. Actually, it would never get formed, for it would straightway fall to pieces. The last number in the list of 'standardized elements' for building atoms is 120. The nuclei (and the atoms as well, naturally) cannot, under any circumstances, have more than 120 protons. It is the number of protons that decisively determines the stability of nuclei to fission. In heavy nuclei the forces of repulsion between protons increase drastically, and the nuclear forces of attraction between distant peripheral particles fall off rapidly. As a result, near the nuclear surface we find raging protons, while the neutrons stand aside. The repulsive forces tear the surface to pieces and the nucleus breaks up into big blocks.

Beta Particles Decay

We have just discussed two models of the atomic nucleus: the shell structure and the liquid drop. Each model has its own particular application. The shell model does a better job when describing the quiescent nucleus that has not been excited by any external causes. The liquid-drop model handles the situation best when the nucleus is under stress, when everything is boiling and the particles are energetically colliding with each other, when they evaporate out and when things get so bad that the nucleus splits to pieces. We've already seen in the Planck theory of quanta that unifying theories stumbled on the wave and corpuscles dual nature of things.

Underlying the generalized theory is the contention that the nucleus behaves in shell-like manner when the numbers of protons and neutrons in it are equal to the magic numbers or close to them. Otherwise, the nucleus behaves like a liquid drop. What is more, this conduct is particularly evident when the number of particles outside the filled and closed shells reaches about two thirds of the number in the succeeding filled shell. It turns out that the particles outside the nuclear filled shells are responsible for all the vagaries of the nucleus, from the ejection of individual particles to the disruption of the whole nucleus. On the inside, the particles in the filled shells behave much more modestly, taking no direct part in these activities of the nucleus. They deform the nuclear surface, as a result, of which the nucleus does not have a spherical distribution of proton charge, and in a number of other nuclear peculiarities.

To sum, nuclear forces act independently from the charge of the particles. They are operative only over short distances, are very strong, and depend on the mutual directions of spin of the interacting particles.

We already know how alpha particles and gamma photons get out of the nucleus. How do beta particles, ordinary electrons, come to be ejected?

Beta radioactivity was problematic since electrons cannot exist in nuclei. Even for exceptionally high speeds of the electron when its energy is of the order of nuclear energies, the length of the de Broglie electron wave is still hundreds of times greater than the dimensions of the nucleus. And the size of the electron cloud, as we have already seen in the case of the Hydrogen atom is of the same order as the wavelength of the electron. **There is no room for the electron in the nucleus** also because its spin, combining with the spins of the nuclear particles, would have produced incorrect values of nuclear spin.

But photons do emerge from the atom despite having much grater wavelength than the atomic dimensions. That might negate against the de Broglie's wave restriction. Then how do electrons come out of the nucleus, when they were never there to begin with? The nucleus has very massive particles that give birth to this extremely light electron. What is more, an electron flying out of a nucleus violates two basic laws of physics: that of the conservation of energy and of angular momentum. Any process in the nucleus such as the ejection of a beta particle can proceed only in such fashion that allows the nucleus to moves from one definite energy level to another one, with the energy difference carried off by the beta particle just as definite. Yet the spectrum of electron energy in beta disintegration did not exhibit even in such quantized energy levels or lines corresponding to definite energies. Further, after the ejection of a beta particle, the nuclear spin remains the same despite of having a new spin carried off by the ejected electron. Further, the nucleus acquires an additional positive charge exactly equal in magnitude to the charge of the released electron. The nucleus becomes ionized.

In a small percentage of beta decay, the mirror images of electrons fly out of the nucleus. They differ from electrons in that they are positively charged. It is even more difficult to account for this type of beta decay. A proton could eject the positron and turn into a neutron. But, unlike a neutron, a proton is absolutely stable with regard to beta decay. This, the *positive beta decay* is less frequent than the ordinary electron-ejection decay.

In science, laws hold for all worlds and all phenomena. Newton's second law of motion requires that motion can neither be created without action nor destroyed without action. One type of action can generate reaction and motion can change form can even become imperceptible. But motion never vanishes. Newton's laws represent the amount of motion in terms of energy and momentum which are the two invariables on non-vanishing motion. Of course, the non-vanishing of motion leads to the more serious question on the very origin of such entity that never vanishes. That question will always remain the bright light at the end of the tunnel that will continue indefinitely. Beta particles possess a continuum of energy from zero to a certain maximum without adhering to any nuclear shell rules.

We are already acquainted with the conversion of protons into neutrons and of neutrons into protons that lies at the heart of nuclear forces. In this process, a proton that has emitted a positive π-meson converts into a neutron, while a neutron converts into a proton upon capturing a meson. The neutron may emit a negative π-meson and convert into a proton. But, in beta disintegration, mesons never leave the nuclei. However, a few years after the discovery of the free neutron, it was found to be unstable particle in contrast to a neutron in the nucleus. A free neutron outside the nucleus converts into a proton in an average of just about 12 minutes after its birth. It is not in this transformation that an electron and a neutrino are emitted. A nuclear neutron should turn into a proton and an electron in quite a different way.

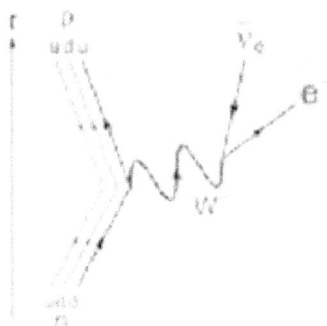
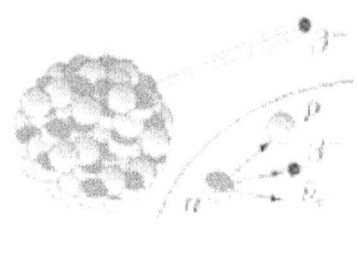

The Feynman diagram for β–decay of a neutron into a proton, electron, and electron antineutrino via an intermediate W boson

β–decay in an atomic nucleus. The intermediate emission of a virtual W boson is omitted.

We recall that nuclei which emit beta particles are either themselves unstable (for instance, the massive nuclei of elements towards the end of the Periodic Table) or are put into an unstable state by neutron bombardment. Thus, the emission of beta particles is initiated during the transition of the nucleus from an unstable state to a more stable state. Nuclear stability is founded on the fact that the protons are the building stones of the nucleus and the neutrons are the cement that holds protons together in a firm structure. When bombarded by a neutron, the nucleus restores equilibrium by converting cement into stones if there is too much, and stones into cement, if they are in excess and threaten the nuclear stability. These transformations take place with the ejection from the nucleus of 'surpluses' or 'deficits' of charge. A proton changes into neutron, gets rid of charge and ejects it in the form of a positron (the positively charged mirror image of the electron). When a neutron turns into a proton, it ejects an electron and thus increases the total charge of the nucleus.

How fast do transformations of nuclear proton and neutron take place?
Not in 12 minutes since a nuclear neutron is different from free neutron. An unstable nucleus cannot stand the unstable state for even thousandths of a second. The nuclear forces that bind the neutrons to the protons could sometimes inhibit the decay of a neutron or proton. Then the nucleus lives a long time before beta decay starts up, a very long time, sometimes as long as hundreds and thousands of years on the average.

At present, Quantum Mechanics cannot predict accurately the mean lifetime of beta-radioactive nuclei. This is due not only to the very approximate knowledge of nuclear architecture (which, in the final analysis, amounts to knowledge of nuclear forces), but also due the unknown nature of the disintegration of a free neutron.

Discovery Of The Neutrino
The violation of the classical and quantum laws by the ejection of beta particles from the nucleus led Wolfgang Pauli to propose a new elementary particle that could preserve the laws of conservation. The conservation of electric charges suggests that the hypothetical particle has neutral charge. The neutron must split into an electron of negative charge and equal and opposite charged particle left behind in the nucleus. The hypothetical particle should have a spin equal to that of the electron, but in the opposite sense. Both spins cancel, yielding zero. Then the spin of the nucleus, when an electron and its accomplice are ejected, remains the same, as required. Finally, the electron and. its mate carry off an energy equal to the maximum energy that electrons are capable of in beta decay of the nucleus. This maximum energy is quantized, that is, it is exactly equal to the difference between the two energy levels of the nucleus prior to and following beta decay. But this energy can be distributed between the electron and its accomplice in any way. The division process is not governed by Quantum Mechanics and no restrictions are imposed on it. Thus, quantization of energy in nuclei is preserved and the laws of conservation of energy and spin are not violated.

The newly proposed accomplice particle should be negligibly small, at least a thousand times less than the electron mass, had no charge; **all it had was energy and spin.** The absence of charge made it look like the neutron, discovered just shortly before. Only it was lighter by a million times. So, it was named '**neutrino**', or little neutron. The neutrino can pass through the entire visible universe-millions of light years-and never give a sign of its existence. It would appear never to interact with anything. The neutrino was finally and definitely established by indirect evidence. The theory of beta decay advanced by Pauli together with the Italian physicist Fermi was unresolved for a quarter of a century.

Electron Capture

In heavy nuclei, the deepest-lying electrons in shells closest to the nucleus find themselves in unstable conditions. On the outside they are repulsed by large numbers of electrons. From the inside they are just as strongly attracted by the heavy positive nucleus. Electron clouds begin to push inwards into the forbidden zone. A slight probability appears of atomic electrons finding themselves in the nucleus. The nucleus captures those falling electrons out of the atomic shell.

In light atoms where the nuclear charge is small and there are few electrons, the prohibitions of Quantum Mechanics are stringently adhered to. The electron 'cloud of probability' does not get into the region occupied by the nucleus.

In the nucleus, the captured electron diminishes the charge of the nucleus by one unit, as in the case of positron beta decay. However, the spin of the nucleus remains the same, even though it received an extra portion from the electron. If that is the case, then the spin brought in by the electron must be carried out of the nucleus by another particle-the *neutrino*.

The difference now is that the neutrino appears when an electron disappears in the nucleus. And so the neutrino is now the only witness to the electron capture by the nucleus. It can collide with another nucleus and convert a proton into a positive electron and a neutron. That is *inverse beta decay*. The invention of nuclear reactors confirmed this scenario. Nuclear reactors produce powerful fluxes of neutrons. When absorbed in the walls of the reactor, they induce artificial radioactivity. Streams of neutrons split up the nuclei of heavy atoms and generate heat and electricity that acne be detected as follows.

The neutrino experiment was performed by Clyde L. Cowan and Frederick Reines in 1956. This experiment confirmed the existence of the antineutrino-a neutrally charged subatomic particle with very low mass. In beta decay the predicted particle, the electron antineutrino ($\bar{\nu}_e$) - should interact with a proton to produce a neutron and positron - the antimatter counterpart of the electron.

$$\bar{\nu}_e + p^+ \rightarrow n + e^+$$

The positron quickly finds an electron, and they annihilate each other. The two resulting gamma rays (γ) are detectable. The neutron can be detected by its capture on an appropriate nucleus, releasing a gamma ray. The coincidence of both events - positron annihilation and neutron capture - gives a unique signature of an antineutrino interaction. Most hydrogen atoms bound in water molecules have a single proton for a nucleus. Those protons serve as a target for the antineutrinos from a reactor. Scintillation counter (filled with a liquid toluene which contained an abundant supply of Hydrogen nuclei) scintillates when gamma photons appear.

To facilitate neutron capture, chemically pure cadmium, whose nuclei avidly absorb free neutrons, is added to the liquid. Here too, neutron capture is accompanied by the emission of gamma photons and produces scintillation. Thus, two scintillations separated by an interval of a millionth fraction of a second should give some clue to the expected phenomenon. These double flashes, which are very infrequent- only a few during many hours of reactor operation, were actually recorded.

CHAPTER 5:
ELEMENTARY NUCLEAR PARTICLES

The world of the elementary particles of matter manifests most clearly the regularities and laws that have been reflected in the wave properties of particles of matter and the material properties of waves.

The slightest deviations in the proportions of protons and neutrons in the nucleus make for instability. At the same time, in order to account for the observed emission from the nucleus during disintegrations of particles that had never been inside it, it was presume that the neutron in the nucleus could convert into a proton, and vice versa. This led to the discovery of the *neutrino*. The stability itself of the nucleus was, it appeared, due to a new particle called the *π-meson*. In the search for this particle, physicists discovered the *μ-meson* (also called muon). Gradually, it dawned on scientists that the world of building stones that make up atoms and their nuclei was not quite as invariable and stable as had been thought.

In the depths of the nucleus there exist unimaginably complex world. In the year 1928, the new world of the elementary nuclear particles was just beginning to unfold. Then, the atom was thought to consist of only the proton and the electron. Quantum Mechanics was only three years old. The Hydrogen atom and the formation of Hydrogen molecules were modeled mathematically with great accord with the spectral analysis, spin, and Stark's effect. The tunnel effect has just helped us to understand the emission of alpha particles by radioactive nuclei. We still know practically nothing about nuclear and other particles. Even the Neutron was not discovered prior to 1932.

We set aside the ether because it did not make sense by virtue of its extreme toughness, invisibility, and its permeation through all matter. Then, in 1905, Einstein invented a space that can be distorted by mass without explaining the mechanism of such effect. Then, in 1928, Dirac advanced another entity of the void that made no better sense than the old ether.

Dirac formulated the Dirac equation and predicted the existence of antimatter. Paul Dirac believed that the successes of Quantum Mechanics may turn out to be short-lived, for this theory grew out of Classical Mechanics which described only relatively slow motions of bodies. In light nuclei, the electron has velocities of the order of thousands of kilometers per second, in the heavier nuclei velocities rise to hundreds of thousands of kilometers per second. Definitely not slow. About twenty years earlier, the special theory of relativity extended Newton's laws into the fast moving objects by tying the time and mass with the velocity of the object. Dirac accounted for the fast motions of nuclear particles by combining it with the special theory of relativity.

The Dirac equation in its original form is as follows:

$$\left(\beta mc^2 + \sum_{k=1}^{3} \alpha_k p_k c \right) \psi(\mathbf{x}, t) = ih \frac{\partial \psi(\mathbf{x}, t)}{\partial t}$$

where
m is the rest-mass of the electron,
c is the speed of light,
p is the momentum, understood to be an operator in the sense of the Schrödinger theory,
$\mathbf{x}$ and t are the space and time coordinates,
$\hbar$ is the reduced Planck constant, h divided by 2π.

Paul Dirac (8 August 1902- 20 October 1984), England.

The new elements in this equation are the 4×4 matrices α_k and β, and the four-component wave function ψ. The matrices are all Herniation and have squares equal to the identity matrix:

$$\alpha_i^2 = \beta^2 = I_4$$

and they all mutually anticommute:

$$\alpha_i \alpha_j + \alpha_j \alpha_i = 0$$
$$\alpha_i \beta + \beta \alpha_i = 0$$

The Dirac equation is superficially similar to the Schrödinger equation for a free massive particle that was written as:

$$-\frac{\hbar^2}{2m}\nabla^2\phi = i\hbar\frac{\partial}{\partial t}\phi.$$

Dirac's purpose in casting this equation was to explain the behavior of the relativistically moving electron, and so to allow the atom to be treated in a manner consistent with relativity. His rather modest hope was that the corrections introduced this way might have bearing on the problem of atomic spectra. Although Dirac's original intentions were satisfied, his equation had far greater implications for the structure of matter, and introduced new mathematical classes of objects that are now essential elements of fundamental physics.

Of course, Dirac's equation constituted greater philosophical exercise than mathematical or physical exactitude. The main reason for resorting to the relativistic treatment of particles was vaguely supported beyond their great speed. But, Dirac's playful manipulation of Schrödinger's wave equation expanded the field for new variables that could explain the very complex behavior of elementary particles.

Relativistic Quantum Mechanics
Newton's Classical Mechanics opened the door for relativistic mechanics by introducing the concept of uniform motion. It implied that objects moving with uniform velocities were essentially at rest since there are no forces acting upon them. Put differently, Newton's laws required that the effect of force must change the velocity of the object. That led to the most crucial search of the twentieth century for a universal frame of reference for time, coordinates, and mass. Essentially, Newton's laws imply that if two objects were ejected at uniform speeds from two unconnected frames of reference, the objects would have no means of relating to each other. Each object would be at rest in its own frame of reference. But, since the two frames are unconnected, then we have no mean to describe their mutual motions had they come in each other's proximity.

Classical Mechanics uses the angular velocity of the Earth around the sun in measuring the time of occurrence of events. Einstein proposed the use of the speed of light as a universal constant. Since the light travels in space at constant speed with no dependency of earthly events. The mystery of the constancy of the speed of light has never been cracked down. What exactly makes photons anchor themselves in an empty void with such exactitude in speed?

The relativistic scheme suggests that the mass of an object increases as it approaches the speed of light due to resisting its accelerating force. There is no force great enough to make an object move with the velocity of light. Thus, the conclusion of the theory of relativity contradicted its underlying axioms. For, there are no means to measure the velocity of an object moving at uniform speed in a frame of reference that is unconnected to the frame where the speed of light in being measured. Thus, the special theory of relativity assumed a fixed universal frame where the speed of light was measured and omitted the fact that the constant speed of light was actually measured in a known frame of reference.

Further, in applying the relativistic transformations, material bodies are assumed those assemblies of particles that can be at rest. Photons cannot be at rest, and so the theory of relativity does not apply to them. This idea is expressed by the famous equation

$$m(v) = m_o / (1 - v^2/c^2)^{1/2}$$

Here, $m(v)$ is the mass the body has when moving with a velocity of v; m_o is the so-called rest-mass which the body has when is not in motion, and c is the velocity of light. As v approaches c, the denominator diminishes, first slowly and then faster and faster. Accordingly, $m(v)$ increases, since m_o is a constant quantity independent of the velocity. Finally, when v is equal to c, the mass of the body, $m(c)$, becomes infinitely great. One must observe that the speed of light, c, is actually measured in the same frame where the mass m_o and time t_o are measured. Thus, the assumption that the speed of light is constant in every place in the universe is disputable.

The paradoxes of the special theory of relativity are apparent from its main conclusion that particles possess rest-masses peculiar to each individual particle. What is the basis for assigning rest-masses such that particle possess such individual identity? What frame of reference would be considered at rest where the rest-mass is measured? For if a particle was traveling at a uniform speed of light, that particle should still be at rest according to Classical Mechanics. Further, the theory postulated that the photons act as corpuscles at very high energy and that fields could pump mass to electrons, while the theory concluded that the mass of the photon is indeterminate. For, by putting the rest-mass of a photon equal to zero in above relativistic relations, we find that for a photon velocity equal to c, the mass $m(c)$ would, be 0/0 or indeterminate. **That means it can have any value whatsoever. That is in total contradiction to the assumption that photons have zero rest-mass. For, if photons could have any value of mass, they should not have zero rest-mass.**

Thus, the special theory of relativity concluded that photons can move only with the velocity of light. To the contrary, no material particle can move at the speed of light. Thus, the velocity of light is an insurmountable barrier between material corpuscles and wave photons. Here, we note the arbitrary and inexplicable conclusion of the theory of relativity that sets the photons in odd with corpuscles. The fact that the **relativistic mass, $m(v)$, could attain infinity is in total conflict with Planck's hypothesis that nothing in physics is boundless or without limit.**

In particle accelerators, charged atomic particles accelerated to high velocities will increase in mass as their velocity approaches that of light. The radius of curvature for a particle moving relativistically in a static magnetic field is

$$r = \gamma\, m\, v\, /\, q\, \boldsymbol{B}$$

where

$$\gamma = \frac{1}{\sqrt{1 - \left(\frac{v}{c}\right)^2}}$$

m is the mass of the particle, q is its charge, $\boldsymbol{B}$ the magnetic field strength, v is its velocity, and r is the radius of its path.

Beside the change of mass of objects moving at higher speeds, changes in velocity that bring bodies close to that of light have other things in store for us. The mass changes and the very course of time itself are altered onto the proper time of the body. In an attempt to interpret the meaning of 'proper time of a body', the theory of relativity reexamined our old system of measurement of time. Relying on the alternation of day and night in measuring time is our subjective time that is geared to our bodily functions. The faster the rhythm of events, the faster the time flows. There is a very definite analogy of this in relativity theory. Relativity theory concluded that the faster a body is moving, the slower its proper time flows so that the body views 'general' time as flowing faster. This is expressed mathematically as follows:

$$t\,(v) = t_o\,(\,1 - v^2\,/c^2)^{\frac{1}{2}}$$

Here, $t\,(v)$ is the time proper of the object's watch, t_o is the time reckoned on an earth clock. From this formula it follows that for photons moving with the velocity of light, time doesn't move at all. Aside from the many paradoxes of the theory of relativity, the equation of the rest-energy will play a very important role in the discovery of new nuclear particles. It is written as follows.

$$E_o = m_o c^2$$

Here, E_o is the energy of a stationary body with rest-mass m_o. To distinguish it from the kinetic or potential energy, we call it the *rest-energy* or the *energy proper* of the body.

It will be seen that it is independent of either the velocity or the position of the body. Classical physics knows only two types of energy. **rest-energy of the body has no place in Classical Mechanics. It is something very special to Quantum Mechanics**. We do not even know whether the concept of rest-mass is without flaws. For **why do material objects not move at uniform speed of light from the instant of the beginning of the universe? or why light choose to have a fixed frame of reference other than ours and to which we have to compare our time, mass, and length?**

Dirac's Quantum Wave Equations

The discoveries of new phenomena in physics motivated Dirac to combine the Schrödinger equation with Einstein's special theory of relativity. Since Schrödinger's wave equation was founded on setting the proper boundary and initial values of problems that could lead to the precise mathematical modeling, Dirac's approach aimed at refining those conditions in hope for greater precision in finding the proper eigenvalues. In fact, Dirac's approach was a desperate effort to accommodate the new discoveries of unexpected particles and nuclear processes.

Dirac postulated that the modified quantum wave equation would yield *relativistically invariant solutions*. Thus, the invariants of total energy and total momentum of Classical Mechanics that preserve the motions of material bodies should remain invariant for the fast moving bodies despite their shrinking proper time and increasing of mass. Dirac's postulate was further supported by the fact that our earthly time is measured in regard to the rotation of the earth about its axis simultaneously with its rotation around the sun. Thus, the earth is always being accelerated and itself accelerates every object on it. Despite the insignificant acceleration of our earthly frame of reference, Dirac's combined equation should eliminate such inhomogeneity in classical motion. It has already been realized that the smallest quantities such as the Planck's constant and the massless photons and neutrinos have played new and crucial roles in Quantum Physics, and therefore, the non-uniform time in our planet and sun was no less important.

The relativistic modification of Schrödinger's wave equation was based on the assumption that the motions of bodies in a spaceship moving with a velocity close to that of light should not differ from those on the earth if, of course, the gravitation is the same. That is, if it has been artificially produced in some way in a homogenous space that is devoid of planets and stars. And since the motions of bodies do not depend on the velocity of the reference system used to reckon their positions in space and time, whether it is the earth or homogenous space, the laws of motion of these bodies must also be independent of the system of reference.

To illustrate, if an equation states that in a spaceship moving at a velocity close to that of light, a material body describes a hyperbola, while on earth it describes a parabola, then the equation is not invariable and violates the very essence of Classical Mechanics. Schrödinger's equation failed the invariability check and compelled Dirac to introduce into the Schrödinger *equation four wave functions* in place of one. The resulting equation was quite unlike the original one. But the new equation yielded excellent relativistically invariant solutions. Dirac's wave equation yielded four 'probabilities' for an electron in place of one. The meaning of the first two solutions would probably have remained obscure for many years if *electron spin* had not been discovered three years before.

The first two solutions of the Dirac equation correspond to the two possible senses of electron spin relative to the direction of motion of the electron. As Ørsted had observed in 1820, the electron is capable of magnetic action. It may be pictured as a minute constant magnet. If an elementary magnet of this kind is introduced into a magnetic field, it will orient itself in this field. In the simplest possible case we have two orientations: one with the magnetic field (absolutely stable), and the other counter to the field (absolutely unstable). When the electron magnet aligns itself with the field, its potential energy in the field is a minimum. In the counter-field case, the energy is a maximum. This can be calculated and converted to a difference of wavelengths of photons emitted in an atom by electrons with spins with and counter to a magnetic field. Then it appears that all doubled spectral lines are split exactly the number of times yielded by the two opposite orientations of electron spin.

The next question is about the 'fat' lines that have split up into three, four and larger numbers of satellites. The two orientations of spin yield only a pair of *satellites*. And there couldn't be any more because the electronic magnets jump straight through to the most stable position. The electron participates in this new motion that produces spin and is also in orbit about the atomic nucleus. A kind of unitary current, and the action it produces resembles that of a small magnet. Thus, an electron in an atom is like a *double magnet*. In place of two orientations, we have three, four and even larger numbers. As the elementary magnet of the electron passes from a less stable orientation to a more stable orientation, it can come to a stop at a number of intermediate positions. The energies of these positions are integral fractions of the maximum energy between the extreme positions of the magnet. These are very definite energies separated by quantum intervals of a specific magnitude. The definite orientation of electron magnets in atoms in a magnetic field is called **'space quantization'**.

A spectral line splits into just as many satellite lines as there are orientations that can be assumed by the electron magnet. Calculations of differences of satellite wavelengths likewise exhibit excellent agreement with experiment. The spin came up so suddenly from Dirac's equation.

Discovery Of Anti-matter

The other two solutions of Dirac's four wave functions are very much alike, just as the first two corresponded to opposite orientations of electron spin. Here, too, we have two opposites: one of the orientations corresponds to positive total electron energy and the other to negative total electron energy. Even though the total energy can have either sign, depending on whether the electron is in free flight or is associated with other particles, as in an atom, yet Dirac's wave equation is written only for a free electron. Thus the negative total energy created a new problem.

Dirac went to look for the origin of the negative total energy. For it might have some meaning. It might be that the negative energy belongs not to an electron but to some other particle with charge opposite to that of the electron. The electron charge is negative, so this particle should have a positive charge. Both should be equal, however, in absolute value. The proton did not fit the description despite its positive charge. The negative energy had to belong to a particle with mass exactly equal to that of the electron. The proton would definitely not do, for it was nearly two thousand times more massive than the electron.

The only possibility was a mirror image of the electron. However, this picture doesn't explain the negative total energy of such a positive particle. If the energy is negative, that means the particle is bound to something. The electron is absolutely free, all other particles, upon solution of the equation, are removed so far away that electrical interaction can be disregarded. The electron is alone in motion in a boundless and absolute void.

Where then do we get this second, positive, particle, and the mirror image of the electron? **Dirac postulated that the void vacuum, which does not contain a single particle except the sole electron, is not empty at all.** Quite the contrary, it is filled to overflowing with electrons.

The positive mirror image of the electron is a **hole** in a filled emptiness. Suppose the empty space contains particles that cannot interact with the instrument. Then even if the space is loaded with particles, we will continue to call it empty space.

Then, how can particles be divested of their ability to interact? Doesn't that contradict the very essence of material bodies? Like electrons residing in atomic levels, in a 'well', they are not able to get out so as to interact with the measuring instrument because they haven't energy enough to interact.

Dirac postulated that the vacuum is completely filled with electrons. The entire universe takes part in the formation of a unified vacuum of infinite extent. Infinite quantities of electrons in it fill infinitely great numbers of energy levels of the vacuum, forming a unified and interrelated assembly of particles. Each energy level in the vacuum is filled according to Pauli's principle such that each level can accommodate two electrons with opposite spins, and no more. The vacuum is thus the common **'universal well'** in which the electrons are located. The potential well's top energy level lies at the energy distance of $m_o c^2$ from the zero of total energy downwards. Thus, all the electrons in the vacuum must have negative energies. **No instruments can detect these vacuum electrons until they jump out of their well.**

Obviously, the empty particle requires energy of $m_o c^2$ to reach the ground level, in addition to another amount of energy $m_o c^2$ that makes the proper rest-mass of the particle. In other words, in order to get out of the vacuum, the electron must not only overcome the barrier of height $m_o c^2$; it also has to acquire the rest-energy $m_o c^2$ to which it is entitled. Therefore, the total height of the barrier separating **vacuum electrons** from their interaction with an instrument is $2m_o c^2$. That is quite a substantial energy. **Supposedly, Dirac accepted [as facts] that particles must have very identifiable constant rest-masses and that light must travel at very identifiable and constant speed.** Further, Dirac reached the conclusion that **energy was the supreme creator of particles from void**.

Why do particles and light were divested with such very identifiable constancies of rest-mass or speed?
Why is energy alone divested with the authority to create elementary particles from void?
No answer is available.

Only after WWII that physicists have been able to impart to electrons energies of this magnitude. When Dirac proposed the idea of filled vacuum and negative energy levels, energies like this were still being dreamed about.

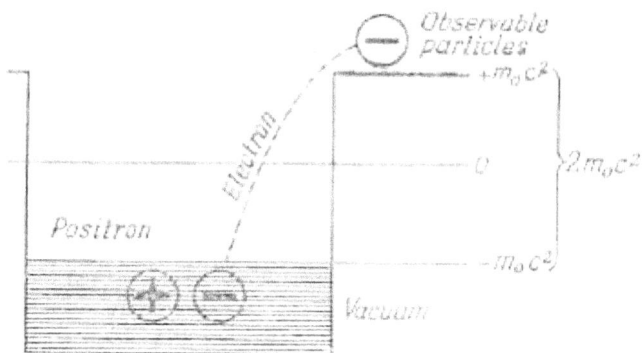

All the energy levels in the vacuum are filled to overflowing with electrons. Electrons exist in the vacuum, but they cannot interact with each other or with any instrument. Their intrinsic spin must comply with Pauli's Exclusion Principle which prohibits more than two electrons to occupy one energy level. These electrons can coexist with us for any length of time and we will never suspect anything, for they never make their whereabouts known in any way. Unlike atomic energy levels, vacuum energy levels are not subjected to the boundary or initial conditions of Coulomb's repulsion and attraction forces and are not subjected to the angular motions around nuclei or the consequence of generating magnetic fields during orbital motions such as the case in atomic shells.

Let us suppose that for some reason an electron in the vacuum acquires enough energy to jump out and becomes a free electron with positive total energy. What happens in the vacuum? **A hole is formed**. The site in the vacuum, where the electron had been, becomes ionized. It gets a positive charge equal in magnitude to the charge of the electron. Now a hole is something we remember from our dealings with semiconductors. There, an electron jumped into the conduction band and left behind a hole in the filled valence band, in which it had negative energy. In semiconductors, the hole is indeed an 'empty site' introduced for the simple convenience of describing different types of electron motion in the valence band and the conduction band. The hole in a vacuum is something quite different. The vacuum hole is in no way different from the electron. It is a real particle, just as real as the electron but with opposite charge.

Like the electron, the hole has a rest-energy of $m_o c^2$, which is an energy equal to the depth of the topmost energy level in the vacuum well. In other words, **the electron and hole originate out of 'nonexistence' only in pairs**. The energy expended in the production of each particle is $m_o c^2$ (their masses are equal), or $2m_o c^2$ altogether. If the free electron meets a hole, both the electron and the hole return to vacuum, or vanish, and become non-existing or non-detectable. Before returning to the vacuum, the electron gives up the energy consumed in ejecting it from the vacuum that was used to generate it and the hole. The energy of **electron-hole annihilation** is carried by gamma photons, which will fly out of the site at which the electron and hole merged.

Why is the electron-hole annihilation energy taken up by gamma photons?
That energy is big enough to correspond to hard gamma rays.
No less than two (and rarely more) gamma photons are generated because the merging electron and hole have opposite spins.
The opposite direction and equal magnitude of the gamma photons conserve the zero total momentum of the electron and hole in the vacuum.
The momenta cancel when two gamma photons emerge and the electron and hole merge..

In 1932, Blackett, together with an Italian scientist, G.P.S. Occhialini, designed the counter-controlled cloud chamber, by which they managed to use cosmic rays incidents as a triggering event of two Geiger-Muller counters to record the interaction. The output impulses of the GM counters turn on the cloud chamber. The two Geiger-Muller tubes were placed one above and one below the vertical the chamber. Blackett and Occhialini exposed a photographic plate to cosmic rays and detected two tracks corresponding to an electron and to an unknown particle of the same mass but with positive charge. The tracks forked out from a single point in different directions. Since the photograph was made in a special chamber placed in a magnetic field, the different directions of the tracks definitely indicated opposite charges.

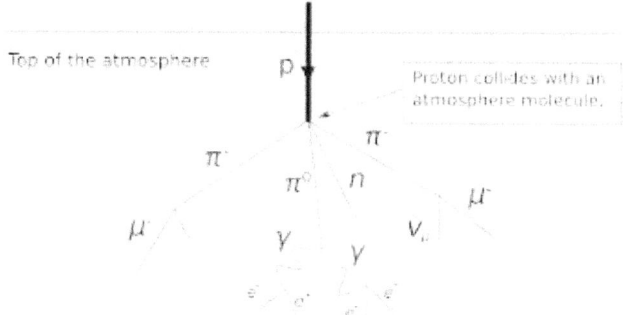

In the spring of 1933 they confirmed Anderson's discovery of the positive electron and demonstrated the existence of "showers" of positive and negative electrons, both in approximately equal numbers. In the interpretation of these experiments Blackett and Occhialini were guided by Dirac's theory of the electron.

.

Giuseppe Occhialini (December 5, 1907 - December 30, 1993), Italy, contributed to the discovery of the pion or π-meson decay in 1947.

Patrick Maynard Stuart Blackett (18th November, 1897- 13 July 1974), England

This fact and the knowledge that positive particles (positrons) do not normally exist as normal constituents of matter on the earth, formed the basis of their conception that gamma rays can transform into two material particles (positrons and electrons), plus a certain amount of kinetic energy - a phenomenon usually called **pair production**. The reverse process - a collision between a positron and an electron in which both are transformed into gamma radiation, so-called **annihilation radiation** - was also verified experimentally.

Thus the hole was recognized and given the name **positron**. Dirac opened the eyes of physicists to utterly new aspects of the invisible world of vacuum. According to Dirac, it is filled with electrons that do not interact with particles in the 'above-vacuum' world. Then an electron leaves the vacuum, a positron is immediately generated. These particles are born and die only in pairs.

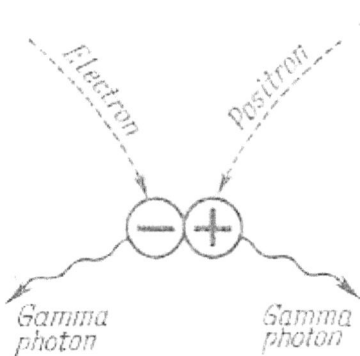

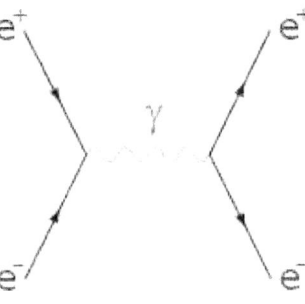

Electron–positron annihilation: $e^- + e^+ \rightarrow \gamma + \gamma$

A Feynman diagram of a positron and an electron annihilating into a virtual photon which then decays back into a positron and an electron via Pair Production.

Dirac's theory considered the vacuum to be filled with particles or antiparticles and that electrons appear only when their anti-particle; the positrons, leave the vacuum. Thus, one particle originates only with the other, its antiparticle.

According to Dirac's theory of pair production and annihilation, the number of electrons and positrons in the world should be the same and nothing could prevent all the electrons from colliding with positrons and falling into the vacuum, leaving behind a disembodied trace in the form of gamma photons. However, an electron rarely meets a positron. So there is no reason to get worried about the world becoming a vacuum. And so, there are more electrons than positrons. But where do they go to? If the electron has its antiparticle, why shouldn't the proton have an antiparticle as well? Generally, every particle should have its antiparticle and, hence, its own vacuum. Then the vacuum should be filled (completely filled.) with neutrons, neutrinos, and mesons, more like a limitless repository (well) for all unborn and dead particles.

Particles can get out of the well only in pairs after acquiring sufficient energy. The first to emerge are, of course, the lightest particles, the neutrinos and electrons. For a proton and antiproton, this energy should be at least two thousand times that for an electron-positron pair. The more massive and sluggish the particle, the more difficult it is to get out of the vacuum. When an electron-positron pair vanishes, energetic photons of gamma rays are born.

Action At A Distance
Dirac's antimatter and vacuum particles created another type of interaction at a distance. In Classical Mechanics, an apple falls to earth due to gravitation. A magnet attracts iron due to magnetic fields. Electrified spheres attract and repel due to electric fields. The earth attracts the moon, and the sun attracts both, although they are all separated by a practical vacuum. The atomic nucleus attracts electrons, although there is an absolute void between them. All this signifies that bodies can interact without any contact. That sort of action is given the name *'field'* to these regions of space in which such action at a distance occurs.

The field was perceived a force acting in 'ether', the extremely tenuous medium that permeated all void. The properties of ether were contradictory on the surface since no mathematician stood in its defense as was the case with Dirac and his filled emptiness. Experiments with light at the end of nineteenth century put the quietus on the ether concept. Within another few years, Einstein's theory of relativity had demonstrated the complete hopelessness of reinstating the ether. **The ether fell, but there was nothing to replace it.** Then, how emptiness could be the carrier of interaction is still beyond the comprehension of man. Surely there can be no doubt that space is the repository of all bodies. A portion of space occupied by matter is called a body, a particle. Then there is space not occupied by any matter. This we call a void, empty space, a vacuum. These two portions are not connected in any way. Emptiness does not act on bodies; neither do bodies interact with the void. True, bodies can interact through empty space, but the void here is of no concern: the interaction is due solely to the bodies themselves.

Albert Einstein tackled the concept of field or action at a distance by the general theory of relativity. It embraces a much broader problem that deals with the relationship of bodies and space such that matter influences the space around it. **Space that is absolutely homogeneous in the absence of bodies loses this homogeneity when a body is 'introduced' into it.**

How does this happen and how do we measure the inhomogeneity?
Einstein replaced Euclidean geometry of empty space that defined the shortest distance between two points as a straight line, with parallel lines never meet, and replaced it with Lobachevsky's geometry. At the beginning of nineteenth century the Russian' geometer Lobachevsky demonstrated that it was possible to construct a geometry just as noncontradictory, internally, as Euclidean geometry, but quite contradictory to common sense, if one gave up the impossibility of intersection of parallel lines. Lobachevsky believed that there is no such thing as a geometry 'in general', applicable to all worlds, that each geometry is determined by the properties of the concrete bodies that it deals with, that the geometry of a space depends on the bodies and things existing in that space and on their configurations. Just as there is no space without bodies, there is no unified homogeneous space. The shortest line between two points in empty space surrounding bodies is now, in the general case, no longer a straight line but a curved line, called a **geodesic**.

The closer the two end-points of our curve are to bodies and the more massive these bodies, the more `curved' is the curve. How do we know? The curvature of space due to bodies is very slight and under ordinary conditions is not noticeable. In order to examine the reliability of the theory of general relativity, researchers took the experiment out into interstellar space and choose some massive object, like our sun, as the 'curving' body. It is naturally most convenient to observe curvature in a line which we consider to be straight. According to Classical Mechanics, the rays from the remote stars must travel in straight lines. This is what Einstein set out to refute.

The experiment pointed a telescope at some star and photographed it. Then, another photograph was taken again when light rays from the same star pass close to the sun. The first picture was taken at night, the second, during a total eclipse of the sun. According to Classical Mechanics, both pictures should show the star in one and the same place on the photographic plate. Whether the light passes near the sun or far away from it should make no difference. According to the general theory of relativity, however, the path of light should curve as it passes close to the sun. The photographic plate should exhibit this curvature as a displacement relative to the first picture of the star.

Nikolai Lobachevsky (1792 -1856), Russia

In August 1919, a special expedition set out for the Arabian Desert to observe a total solar eclipse. One of its tasks was to verify Einstein's prediction. The photographs showed space to be curved; what is more, the curvature was almost exactly as predicted by Einstein. From that time on, physicists' concepts of empty space have changed radically. **Space has become a repository not only of bodies, but also of fields.**

What is a field?
Physicists use this word to describe the space in which bodies manifest interaction. However, there are no noninteracting bodies; all bodies are ultimately made up of particles, none of which are 'indifferent' to the others. For this reason, fields exist everywhere and at all times. And not only between bodies, but within them as well, for there too, there are voids not filled with matter. That is the first and most fundamental property of a field.

From this there immediately follows another conclusion: fields are just as real and universal as is matter. A field is different from matter in one important respect that is: matter is tangible, the field is not (e.g., an electric, nuclear, or gravitational field). But we cannot say that it is impossible to perceive a field. Take an apple falling to the ground. The action of the field is evident from the motion of a body. There is yet another phenomenon displayed by a field-this is light.

As early as 1860, it was established that light is a special so-called electromagnetic field. The wave properties of light were distinct and indisputable. Yet, there was no doubt that light must comprise corpuscles as well. It traveled in absolute void, imparted momentum and carried energy. In 1872, Stoletov found that light is capable of exerting material action by ejecting electrons from metal. In 1900, Lebedev discovered the pressure of light on bodies-just as if light consisted of 'real' particles possessing mass. These two remarkable experiments and the concept of the photon inevitably led to the conclusion that the electromagnetic field has material properties and that field quanta can have the characteristics of particles of matter. Only in 1905, in his theory of the photoelectric effect, Einstein introduced the photon corpuscle. In 1900, the electromagnetic field was quantized, which is to say, it existed in the form of individual particles, quanta of the field. The photons were these field quanta. The history of fields continues to develop.

This was the first span in the bridge across the gap between matter and fields. The de Broglie hypothesis meanwhile was building the bridge from the other end. Electrons could have wave properties. Which meant that mater could behave in a field-like manner. The field, limitless and imponderable, could have dimensions and mass. Matter, limited in space and ponderable, could be deprived of dimensions and mass. The material properties of the field are obvious only at large energies of its quanta. And the field properties of matter come to life only at large energies of its particles. At low energies, the field is then a field, and matter is matter.

There is no Emptiness. The photographically detected joint birth of an electron and a positron is not only the 'opening up' of vacuum. This was the first actual case of a field converting into matter. Confirmation soon came of the reverse prediction of Dirac's theory: the joint annihilation of an electron and a positron upon encounter, and the generation (at the same instant) of two gamma photons. The electron and positron didn't turn into anything, they vanished unchanged into the vacuum. And the energy they released took the form of gamma photons.

Exactly like in an atom where an electron jumps from a higher to a lower energy level releasing its energy in the form of a photon, but at least the electron remains an electron. In the atom, an electron does give up energy, but only a part of it. It can even lose all its kinetic energy in free motion. It can come to a standstill; but the principal energy (the energy proper) is never given up under any circumstances.

For if one gives up an energy $m_o c^2$, which is intimately bound up with the rest-mass m_o, this is tantamount to losing the rest-mass and hence the very essence of being a particle.

We have already observed that particles differ from the quanta of an electromagnetic field in that they can exist at rest and have a mass not equal to zero. This means that when an electron dives into the vacuum giving up its joint positron-electron energy proper, it ceases to be an electron, just as the positron ceases to be a positron. Naturally, their mass does not vanish without a trace, just as their energy does not vanish into nothing. The mass changes its nature and becomes nonmaterial, field-like, while the energy proper converts into the energy of field quanta, gamma photons.

So the vacuum doesn't have any 'real' electrons after all, they are there conceptually, potentially, so to say. The reason is that vacuum or void or emptiness is generally nonexistent. Only matter and fields fill all of space. The vacuum that Dirac had in mind was simply a pictorial image to facilitate depicting the processes of the interconversion of particles of matter and field quanta.

An electron encounters a positron and they convert into gamma-ray photons. If this is possible, then obviously the converse process should be possible: gamma-ray photons converting into particle pairs. This actually does take place provided the photons have sufficient energy, at least $2m_oc^2$. The photons may be observed, recorded and are quite tangible.

The vacuum, on the other hand, is quite intangible until an electron and positron jump out of it. How do we reconcile this situation? Actually, nothing has to be reconciled. Photons are recorded as photons as long as their energy is not great. As soon as it becomes sufficient for the transformation of a pair of photons into a pair of particles, we begin to feel the 'vacuum' properties of the photons.

The photons can vanish with an electron-positron pair taking their place. The term vacuum signifies the possibility of mutual transformations of material particles into field quanta and field quanta into particles.

We have bridged the gap between matter and fields, the traffic can move in both directions: **particles becoming field quanta and field quanta becoming particles**. The important thing is to get up onto the bridge, which is rather high-the energy height measures $2m_oc^2$ which for electrons signifies millions of electronvolts and for protons, thousands of millions of electron-volts. It is convenient to picture vacuum as a universal sea with dolphin-like particles jumping in and out.

Field And Matter
This duality of wave and matter exists everywhere and at all times. The wave has indeterminate extension and eternal mobility. It is clearly a field entity. Further, the de Broglie hypothesis in effect implies that material particles have field properties. It supplements the Einstein hypothesis, which imposes material properties on the field quanta (photons). But, what is mobility or extension if there was no matter in the empty void? We have seen that Quantum Mechanics transformed the world of elementary particles into pure mathematical sport. New physical constants were postulated with minuses and pluses attached to masses, charges, spins, parities, and energies in the stringent and precise manner that defies the crude and objective laws of Classical Mechanics. Our subjective perception of matter was pushed aside by Quantum Mechanics in order to accommodate the inaccessible inner nuclear system and its governing laws. Only inference and deduction governed the efforts of physicists in interpreting the new experimental data regarding the elementary particles.

How do the field properties of elementary particles manifest themselves?
We have already encountered numerous instances. Most typical of these properties is the fogginess of electrons and other particles in space. Particles are nonlocalized. An electron is here, yet it is not here. In attempts to measure its velocity of motion with exactitude we cease to be able to say anything of its whereabouts. This is very typical of a field; it is impossible to localize a field due to its being 'everywhere'.

If we increase the velocity of an electron, it becomes heavier as we approach the velocity of light. Where does it get its extra mass? Electrons are usually accelerated by electric fields. During acceleration, the electric field enters the electron, imparts to it a portion of its energy. Since the energy of the electron is increasing, Einstein's relation implies that the electron velocity and mass should also increase.
$m(v) = m_o / (1 - v^2/c^2)^{1/2}$
But this process of 'pumping' mass from the field into the particle cannot go on endlessly. The mass builds up with extreme rapidity and finally the kinetic energy of the particle becomes equal to its energy proper. This occurs when the velocity of the particle reaches about 80 per cent of that of speed of light. At this point, a new process sets in, in which the wave, or field-like, properties of the particles dominate.

The particles are now in a position to rid themselves of the accumulated energy and their proper energy and to release quanta back into of the field. The increased mass of the accelerated particles requires greater electric field in order to remain accelerated. If the electric field is not increased, the heavier electrons cease to acquire more velocity and thus resist any build-up of energy. The resistance increases as the electrons approach transformation into the field. **Particles**

can never move with the velocity of propagation of the field, and the field can never propagate with a different velocity.

Decay of Subatomic Particles

Up till now, particle transformations have dealt only with the electron (and, of course, the positron). After the discovery of the neutron, it was found that it too is capable of transformation, but, unlike the electron, not into field quanta but into other particles. Outside the nucleus, free neutrons are unstable and have half-life of 613.9 ± 0.8 s (about 10 minutes, 14 seconds). Free neutrons (n^0) decay by emission of an electron (e^-) and an electron antineutrino (v^c) to become a proton, a process known as *beta decay*:

$$n^0 \rightarrow p^+ + e^- + v^c$$

In the nucleus of the atom, the instability of a single neutron to beta decay is balanced against the instability that would be acquired by the nucleus as a whole. Inside the nucleus, protons participate in repulsive interactions with the other protons that are already present in the nucleus. As such the neutrons inside the nucleus are bound and are more stable that the free neutrons. The bound neutron is converted into a proton (p^+) and a π-meson (π^0, π^+, and π^-). It was found out later on that the second transformation of the neutron is not very different from the first. A free π-meson decays into a μ-meson (μ^-) (which is roughly one-fourth as heavy) and a neutrino. In turn, the μ-meson decays into an electron, a neutrino and an antineutrino as follows:

(1) neutron à proton + π-meson à μ-meson +neutrino

$$n^0 \rightarrow p^+ + \pi^- \rightarrow \mu^- + v^c$$

(2) μ-meson à electron +neutrino

$$\mu^- \rightarrow e^- + \bar{\nu}_e + \nu_\mu, \quad \mu^+ \rightarrow e^+ + \nu_e + \bar{\nu}_\mu$$

The same reasoning explains why protons, which are stable in empty space, may transform into neutrons when bound inside of a nucleus. A bound proton can also transform into a neutron via **inverse beta decay**. This transformation occurs by emission of a positron (also called anti-electron) and a neutrino:

$$p^+ \rightarrow n^0 + e^+ + v^c$$

Positron capture by neutrons in nuclei that contain an excess of neutrons is also possible, but is hindered because positrons are repelled by the nucleus, and quickly annihilate when they encounter electrons.

The **μ-mesons** or **muon** is an elementary particle similar to the electron, with a unitary negative electric charge and a spin of ½. Together with the electron, the **tau-mesons** and the three neutrinos, it is classified as a **lepton**. It is an unstable subatomic particle with a mean lifetime of (2.2 μs), compared to that of a free neutron (~12 minutes), a free proton at 6.6×10^{33} years, and that of an electron, with a half-life at least as great as that of the universe, and possibly infinite.

Like all elementary particles, the muon has a corresponding antiparticle of opposite charge but equal mass and spin: the **anti-muon** (also called a *positive muon*). Muons are denoted by μ^- and antimuons by μ^+. Muons were previously called **μ-mesons**, but are not classified as mesons by modern particle physicists. The dominant muon decay mode is the simplest possible. The muon decays to an electron, an electron-antineutrino, and a muon-neutrino. Antimuons, in mirror fashion, most often decay to the corresponding antiparticles: a positron, an electron-neutrino, and a muon-antineutrino. These two decays are written as follows:

$$\mu^- \rightarrow e^- + \bar{\nu}_e + \nu_\mu, \quad \mu^+ \rightarrow e^+ + \nu_e + \bar{\nu}_\mu.$$

The difference between particle decay inside and outside the nucleus is attributed to the different electric forces and the much more powerful nuclear forces inside the nucleus that make the nucleus stable. If there is a new type of force, that means there is a new field or quanta. The carriers of electromagnetic interactions are photons. By analogy, **the carriers of nuclear interactions must be π-mesons** we have already mentioned that μ-mesons interact with nuclei weakly and therefore cannot be quanta of the nuclear field). Unlike photons, π-mesons have a rest-mass, which is rather substantial in the world of subatomic things, for it is nearly three hundred times more massive than the electron. For this reason, π-mesons cannot move with the velocity of light.

The harmonious and proportioned picture of the interrelationships of fields and matter that physicists had just described suddenly broke down. π-mesons turned out to be the very limit of duality. They are matter in that they have a nonzero rest-mass, they represent a field in that their **spin is zero**.

A **π-meson** or **pion** (denoted by π) is any of three subatomic particles: π^0, π^+, and π^-. Pions are the lightest mesons and they play an important role in explaining the low-energy properties of the strong nuclear force. The $\pi^{\pm}$ mesons have a mass of 139.6 MeV/c^2 and a mean lifetime of $2.197034(21)\times10^{-8}$ second. They decay due to the weak interaction with probability 0.999877, into a muon and a muon neutrino as follows.

$$\pi^+ \longrightarrow \mu^+ + \nu_\mu$$
$$\pi^- \longrightarrow \mu^- + \nu_\mu$$

The second most common decay mode of a pion, with probability 0.000123, into an electron and the corresponding electron neutrino was discovered at CERN in 1958 and is given by:

$$\pi^+ \longrightarrow e^+ + \nu_e$$
$$\pi^- \longrightarrow e^- + \nu_e$$

After the rise of Quantum Mechanics physicists established yet another sharp distinction between the spin of particles of matter and field quanta. It was found that **'true' particles of matter can have only a spin equal to one-half the modified Planck constant $h/2\pi$**, whereas **field quanta must have spin equal to zero or to an integral number of** $h/2\pi$. It was found that the magnitude of spin exerts an essential influence on the behavior of microentities which was postulated by the Pauli's Exclusion Principle. It requires that no two electrons in an assembly can exist in exactly the same states. This goes not only for electrons, but for protons, neutrons, and generally any particles with half-spin. Thus, particles with spin zero or with integral spin will violate Pauli's principle. Photons do not have rest-mass and thus need not comply with Pauli's Exclusion Principle. There can be any number of photons in the same states, that is, with the same frequency and the same direction of spin because photon's spin is equal to unity.

Incidentally, it followed from this division of spins that the μ-meson that physicists first stumbled over could not be the quantum of the nuclear field. It has half-spin. **But the π-mesons all have zero spin and hence can serve as field quanta except that π-mesons have nonzero rest-mass.**

Maybe the neutron is simply a compressed combination of a proton and a π-meson?
The rest-mass of the neutron and that of the proton are, respectively, approximately equal to 1,839 and 1,836 electron masses, while the rest-mass of the electrically charged π-meson is 273. Thus, when a neutron emits a π-meson, the neutron should reduce by 273 electron masses and not 3. Therefore, **neutrons are not composites of protons and π-mesons.**

When a free neutron disintegrates, this problem does not arise. The neutron loses an electron- one electron mass. In addition, it imparts to the electron and neutrino a double proper energy of the electron, after which it acquires the mass of the proton.

If a π-meson is emitted, the neutron should lose nearly a hundred times more mass but neutrons are not reduced after the birth of a π-meson. The prohibition is due to the fact that no particle can have mass less than the rest-mass. To emit a meson it would have to find some place in its interior the equivalent of 270 electron masses. This is in direct violation of the law of conservation of energy. The secret of the neutron decay inside the nucleus remains a mystery to Classical Mechanics.

Can a meson be exchanged between a proton and a neutron along the lines of the electron exchange in the Hydrogen molecule?
This would be much simpler, since electrons do not experience any kind of transformations yet a bond is established between the atoms. Maybe a negative π-meson could circulate about two protons? Scientists recently succeeded in binding μ-meson in the atomic cloud in place of an electron, and the μ-meson did the job just as well. The μ-meson joined two atoms of Hydrogen into a single molecule (just like an electron does), the so-called **meso-molecule of Hydrogen**. Since the μ-meson is approximately two hundred times more massive than the electron, its cloud of probability is just that much closer to the nucleus, which means that the μ-meson holds two atoms into a molecule 200 times smaller in size.

However, this is not a π-meson, but a μ-meson. And, again, the forces operative in the meso-molecule are not nuclear forces but electrical forces. The latter are much weaker than nuclear forces. A π-meson cannot hold its place in the atom like all electrons because it strongly and very specifically interacts with the nucleus. The π-meson converts a neutron into a proton, and a proton into a neutron through mysterious routes.

The π-mesons are involved in the emission of a meson by one particle and its capture by another. These processes of emission and capture conflict with the laws of conservation. Yet, the processes exist similar to the tunnel effect of elementary particles penetrating through potential barriers. The Schrödinger equation demonstrated the probability that a particle in a well could get outside without acquiring any energy at all. This too seems to conflict with the law of conservation of energy. In the classical sense, we have a breach of the law of conservation of energy. But Quantum Mechanics permits a probability of seeping through a potential barrier without violation of the law. In the very same manner we can explain the emission and absorption of π-mesons by nuclear particles. Also, the Heisenberg relation between energy and time may be applied to the energy proper of the particle in such fashion that the neuron energy during the emission of π-meson was undetermined during the short time of interaction.

Thus, the thinning out of a neutron that has emitted a negative π-meson, or the loss of a proton upon ejection of a positive π-meson, and also the 'fattening' of particles that have absorbed mesons may be regarded as an uncertainty in the energy proper of these particles associated with an uncertainty in their masses. It is clear that this uncertainty is no less in magnitude than the energy proper $\Delta E = m_\pi c^2$ of the π-meson, where m_π is the rest-mass of the π-meson. From this relation, Heisenberg's Uncertainty Principle yields the uncertainty of the time in which ΔE can exist. It determines the duration of the shuffling of π-meson between the nuclear neutron and proton. From the Heisenberg relation, we get $\Delta E \times \Delta t \approx h$. Thus
$$\Delta t \approx h / m_\pi c^2$$
Putting into this relation the values of mass of the π-meson m_π the Planck constant h and the velocity of light c, we get $\Delta t \approx 10^{-23}$ second. Rather a short time. What distance can the π-meson cover in this time? There is obviously a limit: the π-meson can have a velocity only less than that of light. Therefore, the limiting distance covered by a π-meson from the nuclear particle that emits it is $R \approx c \times \Delta t \approx 3 \times 10^{10} \times 10^{-23} = 10^{-12}$ cm. This coincides, in order of magnitude, with the range of nuclear forces.

Thus, to detect π-mesons during ejection or absorption in exchange from nuclear particles is impossible within the ultimately small $\Delta t \approx 10^{-23}$ sec for the same reason that we cannot detect electrons during their passage under potential barriers.

Underlying the nuclear interactions are the wave properties of elementary particles. Nuclear forces have a limited range of action for the very reason that the nuclear-field quanta (π-mesons) have a nonzero rest-mass. The π-meson exhibits stable conduct only in the nucleus. In the free state, this particle behaves quite differently. Once outside the nucleus, the π-meson decays in a very short time. A positive π-meson converts into a positive μ-meson. A negative one changes into a μ-meson of the same sign of charge.

$$\pi^+ \rightarrow \mu^+ + \nu_\mu$$
$$\pi^- \rightarrow \mu^- + \nu_\mu$$

During decay, a neutrino is ejected. Somewhat later a third π-meson was discovered -an electrically neutral particle. This **zero-π-meson** decays a thousand million times faster than other mesons. Its decay gives birth to two gamma-ray photons, but of far greater energy than those produced in electron-positron encounters. The π^0 meson has a slightly smaller mass of 135.0 MeV/c^2 and a much shorter mean lifetime of 8.4×10^{-17} second. This pion decays in an electromagnetic force process. The main decay mode of zero-π-meson, with probability 0.98798, is into two gamma ray photons as follows:

$$\pi^0 \rightarrow 2\gamma$$

Its second most common decay mode, with probability 0.01198, into a photon and an electron–positron pair is as follows:

$$\pi^0 \rightarrow \gamma + e^- + e^+$$

The rate at which pions decay is a prominent quantity in many sub-fields of particle physics. This rate is parameterized by the pion decay constant (f_π), which is about 90 MeV. It is this instability of π-mesons that makes them so different from photons. Photons can change their energy and can even vanish completely in particles giving up their energy. **The π-meson is certainly the biggest hybrid of particle and quantum** yet.

Deduction Of Coulomb's Law

The electromagnetic field is merely 211 years old. The first observation that an electric current creates a magnetic field was made by Ørsted on July 1820. A year latter, in 1921, Michael Faraday invented the electric motor by relying on Ørsted's induced magnetism.

That led to the observation that a wire moving in the field of a permanent magnet will have a flow of electricity in proportion to the strength of the magnetic field and the speed of rotation. This force is known as the **Lorentz force**, and is given by

$$\mathbf{F} = q\mathbf{v} \times \mathbf{B}.$$

where **F** is the force, q is the electric charge of the particle, **v** is the instantaneous velocity of the particle, and **B** is the magnetic field, in teslas.

The electric field is produced both by stationary and moving charges, the magnetic field, only by moving charges. Since every interaction of charged particles is associated with motion and is manifested in motion, it may generally be stated that every interaction involves the composite electromagnetic field. In the case of a stationary electric charge, an electrostatic field is generated such as like charges repulse, unlike charges attract. Thus, the field induced force with specific intensity and direction creating lines of force.

The two kinds of electric charge- positive and negative- correspond to protons with positive charge and electrons negative charge. The antiparticles were discovered after 1932, and both particles and antiparticles are the only absolutely stable carriers of charge.

In order to explain the creation of photons from electrons, Dirac's vacuum antiparticles are combined with Heisenberg's uncertainty principle as a theoretical basis for deciphering the transmutation of filed to matter and vice versa.

All negative charges belong to electrons. Each electron curves around other electrons due to the inhomogeneity of the space created by their material content. This curvature, however, is due to the mass, not the charge. Accordingly, we have a different field, the field of gravitation. The electron in the interacting part of the vacuum repulses the vacuum electrons that cannot be detected due to their negative rest-energy.

Thus, the interacting real electrons, each should act on the vacuum electrons in similar fashion. But the repulsion of the real electrons and the vacuum electrons is mutual. The vacuum electrons will do the same with respect to the second electron. This will find expression in the mutual repulsion of both real electrons. But, how do the undetected vacuum electrons interact with real electrons? As we stated earlier, in the mathematical sport of the subatomic world, signs and imaginary numbers supplant our subjective

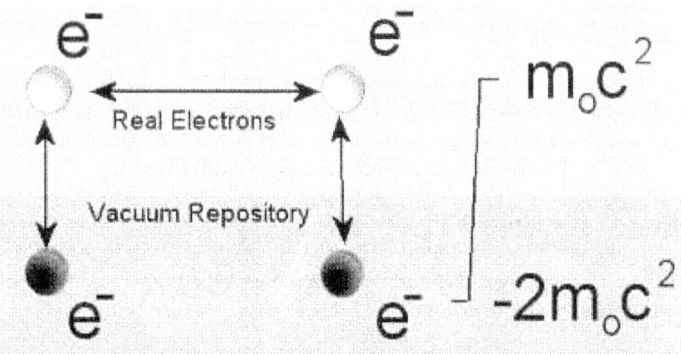

views. We are no longer dealing with sensual classical objects that could be seen, touched, heard, or smelled with direct human senses or crude instruments. Our only guide in such uncharted territory of inaccessibly ultra-small space and ultra-short time processes is the mathematical equations and mathematical logic.

What is the vacuum addition to the energy of the Hydrogen electron? If we use the Planck relation and convert it into frequency, then it will come out not among the gamma rays or even those of visible light, but in the high-frequency radiowaves band which could not be discovered by conventional spectral methods.

After World War II, when high-frequency radio oscillators were built and Hydrogen atoms were irradiated with high frequencies, they immediately responded to the frequency that fits the vacuum addition. A deep dip appeared at the site of .this frequency in the 'radiospectrum' of Hydrogen-the Hydrogen electron was actively absorbing quanta on this frequency. A little while later a second vacuum effect was discovered.

The concept of particle interaction explains that a free and stationary electron can spontaneously emit photons. In the process the electron changes its energy state. But if the absorption and emission of photons by free electrons is reversed instantaneously, the energy of the electron will remain the same, Quantum mechanics allows such processes within the framework of the Heisenberg's Uncertainty relation. The speed with which an electron emits a photon and captures it again should depend solely on the photon energy. The greater the energy of the photon, the faster the electron will complete the operation.

The electron can emit photons of any energy and such photons can move away from their parent to any distance conceivable. However, photons of a very definite frequency have range of action is of the order of the wavelength of the photon. In visible light, this distance is of the order of a fraction of a micron. Photons emitted by different electrons could interact and never return to their parent electrons. They may be absorbed by a distant electron. The electron energy will change by exactly as much as is contained in the non-returning photons, and both electrons will move apart. The farther electrons are separated from each other, the smaller the energy of their interaction.

In the process, the total energy of the photons and electrons remains unchanged, just exactly what it was in the beginning. But of course there never was any beginning or end of the interaction of two electrons. There is no turning off or on of interaction. No matter how far away the electrons are from each other, there will always be some kind of interaction between them. The field of the emitting electron is somehow attached to it despite the fact that photons are independent-acting entities. Very energetic photons emitted by an electron can, during the time of its very short permitted life, convert into an electron-positron pair. Thus, in place of one electron we have two electrons and a positron.

Everything takes place in too short a time. A simple calculation with the Heisenberg's relation shows that our instant lasts about 10^{-21} second. During this time, the photon was able to give birth to a pair consisting of a second electron and a positron at a distance of about 10^{-11} centimeter from the first electron. This is exactly the magnitude that is characteristic of the smallest fogginess of an electron in space. 10^{-11} cm is the length of a de Broglie electron wave moving with a velocity close to that of light. This shows that at the very heart of the wave properties of matter is interaction-the field of the electron. The electron is clouded because it dives into the vacuum and emerges from it near the same site numberless times every second. This postulated electron behavior is called **'trembling electron'**. In this process, an electron can oscillate while located at any place within the locality appropriated for it. This locality is determined by the energy and, hence, the wavelength of the photons that can generate electron-positron pairs.

A photon emitted by some electron may be captured by a different electron.. But electrons are all alike, and there is no way of finding out which one captured the emitted photon. Electrons strive to get away from each other as far as possible. But even when the distance between them exceeds many times over the degree of their 'vacuum fogginess', photons catch up with them and push them apart still more. The greater this distance, the less the energetic photons can overcome it, which means that less energy will be imparted to the electrons in photon exchange and the electron repulsion will be more feeble. **That explains Coulomb's law.**

Electronic interaction is all-pervasive. All the electrons in the universe participate. We might say that the boundless electromagnetic field is found in every corner of the infinite world. Two particles with opposite charges and the same masses, mirror images, meet as they jump out of the mirror and cancel their charges and convert into quanta of the field that performs the interaction.

Virtual vacuum can suddenly become very real. We have already spoken about the two electron magnets. One of them was due to the motion of the electron about the atomic nucleus, the other was caused by the spin motion of the electron. In a magnetic field, these two magnets combine into a sort of unified magnet of a definite magnitude. Yet, **the measured sum of the two magnets was greater that sum of the individuals added together.** The addition in the magnitude of

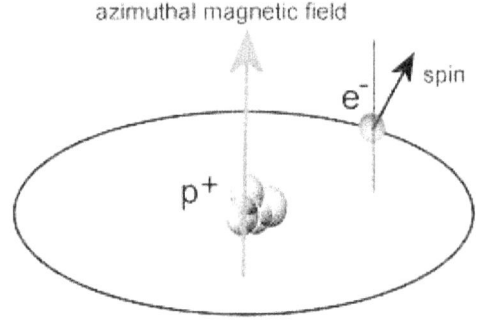

the magnet is due to the interaction of electron and vacuum. An electron moving in an atom repulses the vacuum electrons all along its path like a stationary ship only displaces water, while a moving vessel makes the water move as well.

The transfer of motion from the electron to the vacuum produces in the latter a **current of vacuum electrons.** The magnetic effects of virtual current are added to those that correspond to the motion of the 'real' electron.

Classification Of Elementary Particles

The subatomic universe is governed by unprecedented laws and interrelations among particles and fields. New particles are being discovered with new peculiarities and newer advancement on the path of knowledge. The cosmic rays were the only suppliers of new particles due to their enormous energy. New instruments were invented, old instruments were refined and fresh expeditions set out to mountain peaks, up into the clear air closer to the sky, where exotic nuclear reactions take place. Nearly every year saw the discovery of dozens of new particles. π-mesons were the first.

At the beginning of the fifties, particles were discovered that were more massive than protons and neutrons. Those were called **hyperons**. Cosmic radiation presented physicists with a very valuable gift, a group of K-mesons. Then, a series of gigantic particle accelerators were put into operation that excelled protons to close-optical velocities. Two new particles were discovered that confirmed the predictions of Dirac's theory. They were the **antiproton** and the **antineutron**. Today, there are about thirty different types-about 25 years ago there were only four.

In reviewing the newly discovered elementary particles, the first thing one notices is the broad range of masses: from the electron mass to two and a half thousand electron masses in the case of the **xi-hyperon**. In that range of masses, the distribution of particles as to mass is rather uneven. They come in groups of two and three with similar masses. As for the charges and spins of the new particles, **the charges and spins don't show any sign of diversity**. Charges of the new elementary particles can have three values: +1, 0 and -1, where -1 is the charge of the electron. The spins have three values as well: 1, 1/2, and 0 in Planck units $h/2\pi$. Finally, most of the particles in the updates list of elementary particles are unstable. On the average they have lifetimes ranging from millionths of a second (μ-mesons) to thousands of millions of times smaller fractions of a second (π-zero-mesons). These two lifetimes are the extremes. In the middle of the range are unstable particles with lifetimes from hundred millionths to ten-thousand millionths of a second.

As an illustration, take the positron. It is stable in the sense that it does not decay into any other particles. Yet it doesn't live long in our world-as soon as it meets an electron, it vanishes, as a rule. On the other hand, π-mesons which are unstable in the free state never decay within nuclei. Electrons and neutrinos most often appear in the decay products of unstable mesons and the neutron.

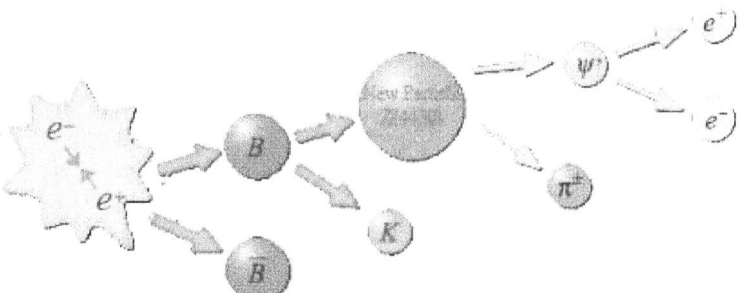

The following grouping of the newly discovered elementary particles is listed together with the old constituents of the atom.

118

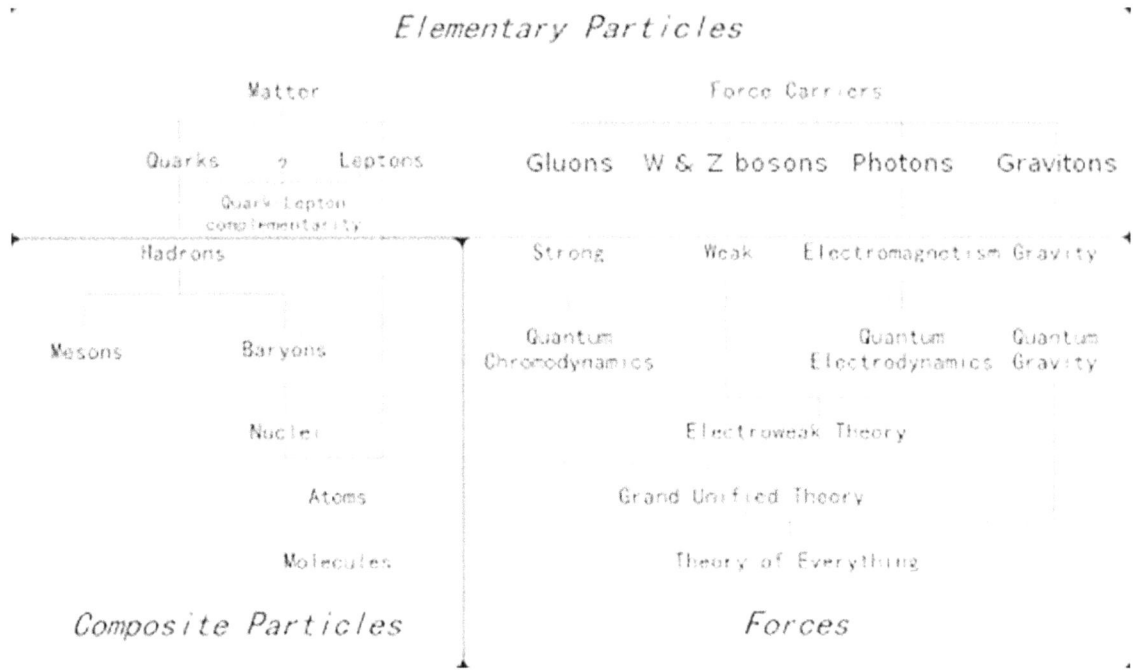

Subatomic Particles in Physics

Class of particle	Name	Designation	Mass (in electron masses)	Charge	Spin (in Planck units $h/2\pi$)	Lifetime in seconds	Decay scheme
Leptons (light particles)	Photon	γ	0	0	1	Stable	
	Electron	e^-	1	−1	1/2	ditto	
	Positron	e^+	1	+1	1/2	ditto	
	Neutrinos 1 and 2	ν	0	0	1/2	ditto	
	Antineutrinos 1 and 2	$\bar{\nu}$	0	0	1/2	ditto	
	Mu-minus-meson	μ^-	206.7	−1	1/2	2.2×10^{-6}	$\mu^- \to e^- + \nu + \bar{\nu}$
	Mu-plus-meson	μ^+	206.7	+1	1/2	2.2×10^{-6}	$\mu^+ \to e^+ + \nu + \bar{\nu}$
Mesons (medium particles)	Pi-minus	π^-	273.2	−1	0	2.6×10^{-8}	$\pi^- \to \mu^- + \bar{\nu}$
	Pi-plus	π^+	273.2	+1	0	2.6×10^{-8}	$\pi^+ \to \mu^+ + \nu$
	Pi-zero	π^0	264.2	0	0	2.2×10^{-16}	$\pi^0 \to 2\gamma$
	K-minus	k^-	966.5	−1	0	1.2×10^{-8}	$k^- \to 2\pi^- + \pi^+$ or $2\pi^0 + \pi^-$

Class of particle	Name	Designation	Mass (in electron masses)	Charge	Spin (in Planck units h 2π)	Lifetime in seconds	Decay scheme
Mesons (medium particles)	K-plus	K^+	966.5	+1	0	1.2×10^{-8}	$K^+ \to 2\pi^+ + \pi^-$ or $2\pi^0 + \pi^+$
	K-zero	K^0	974.2	0	0 $\{K^0_1$	1.0×10^{-10}	$K^0_1 \to \pi^+ + \pi^-$ or $2\pi^0$
	Anti-K-zero	$\bar{K}^0$	974.2	0	0 $\{K^0_2$	6.1×10^{-8}	$K^0_2 \to 3\pi^0$

	Class of particle	Name	Designation	Mass (in electron masses)	Charge	Spin (in Planck units h 2π)	Lifetime in seconds	Decay scheme
Baryons (heavy particles)	Nucleons (nuclear particles)	Proton	p	1,836.12	+1	1/2	Stable	
		Antiproton	$\bar{p}$	1,836.12	-1	1/2	ditto	
		Neutron	n	1,838.5	0	1/2	1.0×10^3	$n \to p + e^- + \nu$
		Antineutron	$\bar{n}$	1,838.5	0	1/2	1.0×10^3	$\bar{n} \to \bar{p} + e^+ + \nu$

	Class of particle	Name	Designation	Mass (in electron masses)	Charge	Spin (in Planck units h 2π)	Lifetime in seconds	Decay scheme
Baryons (heavy particles)	Hyperons (large particles)	Lambda-zero	Λ^0	2,182.8	0	1/2	2.5×10^{-10}	$\Lambda^0 \to p + \pi^-$ or $n + \pi^0$
		Anti-lambda-zero	$\bar{\Lambda}^0$	2,182.8	0	1/2	2.5×10^{-10}	$\bar{\Lambda} \to \bar{p} + \pi^+$ or $\bar{n} + \pi^0$
		Sigma-plus	Σ^+	2,327.7	+1	1/2	8.1×10^{-11}	$\Sigma^+ \to n + \pi^+$ or $p + \pi^0$
		Anti-sigma-plus	$\bar{\Sigma}^+$	2,327.7	-1	1/2	8.1×10^{-11}	$\bar{\Sigma}^+ \to \bar{n} + \pi^-$ or $\bar{p} + \pi^0$
		Sigma-zero	Σ^0	2,331.8	0	1/2	$<10^{-11}$	$\Sigma^0 \to \Lambda^0 + \gamma$
		Anti-sigma-zero	$\bar{\Sigma}^0$	2,331.8	0	1/2	$<10^{-11}$	$\bar{\Sigma}^0 \to \bar{\Lambda}^0 + \gamma$
		Sigma-minus	Σ^-	2,340.6	-1	1/2	1.6×10^{-10}	$\Sigma^- \to n + \pi^-$
		Anti-sigma-minus	$\bar{\Sigma}^-$	2,340.6	+1	1/2	1.6×10^{-10}	$\bar{\Sigma}^- \to \bar{n} + \pi^+$
		Xi-zero	Ξ^0	2,565	0	1/2	1.5×10^{-10}	$\Xi^0 \to \Lambda^0 + \pi^0$
		Anti-xi-zero	$\bar{\Xi}^0$	2,565	0	1/2	1.5×10^{-10}	$\bar{\Xi}^0 \to \bar{\Lambda}^0 + \pi^0$
		Xi-minus	Ξ^-	2,580.2	-1	1/2	1.2×10^{-10}	$\Xi^- \to \Lambda^0 + \pi^-$
		Anti-xi-minus	$\bar{\Xi}^-$	2,580.2	+1	1/2	1.2×10^{-10}	$\bar{\Xi}^- \to \Lambda^0 + \pi^+$

The newly discovered elementary particles put the Quantum Mechanics in a tough spot. New laws were introduced haphazardly whenever a new peculiar behavior was observed in experiment. The following questions will guide the future course of Quantum mechanics:

(1) What decay schemes affect each elementary particle?
(2) Why do particles choose only one or two of a large variety of possible decay schemes?
(3) What are the underlying bases for the great diversity in particle masses?
(4) What is the limiting mass for elementary particles?
(5) Why do particles exhibit groups of closely related masses of two, three and four particles?
(6) Why does the charge of particles have only three values and the spin, two?
(7) Why are most particles unstable, while some are stable?

The following is a brief attempt to refine the above questions.

The particle masses in one group are very close to one another if compared to the broad interval that separates one group from the next. This has been accounted for in an interesting way: a group of particles is actually only a single particle that appears in different guises. For example, the masses of the π-minus- and π-plus-mesons are equal and differ from the mass of the third, electrically neutral, π-zero-meson. Maybe the higher mass of the charged particles is due to their having charge.

We have already mentioned the fact that the field accounts for a portion of the mass of a particle. Since π-mesons are the quanta of the nuclear field, and this field is very much stronger than the electromagnetic field, it would be reasonable to suppose that the bulk of the mass of π-mesons is due to the **nuclear field**, while any addition to it of the electromagnetic field (associated with the presence of charges) would make only a small contribution.

For this reason, charged π-mesons are more massive than the neutral particle, which naturally should be of nuclear origin entirely. This would likewise appear to account for the fact that **lightweight particles do not form triplets. The nuclear field differs in that its quanta have nonzero rest-mass, whereas the quanta of the electromagnetic field are photons with zero rest-mass.**

The electron and positron and both μ-mesons are of a pronounced non-nuclear, electromagnetic origin. Thus, they have no neutral particle, only a positive and a negative particle, a **doublet**.

This doesn't work for K-mesons. The neutral K-meson is more massive than the charged particles. Here the electromagnetic field would seem to be 'subtracted' from the nuclear field. **Hyperons, which are of nuclear origin and have no triplets**, seem to confirm this.

Up until 1955, the nucleon group consisted only of the proton and the neutron. This was quite a team: a doublet made up of a charged particle and a neutral particle. The mystery was resolved, so it appeared, when the negatively charged **anti-proton** was discovered, for here was now a normal triplet like the group of π-mesons. However, the neutral neutron was heavier, not lighter, than the proton and its antiparticle. Again the electromagnetic field appeared to be 'subtracted' from the nuclear field. Most important, however, was the fact that the proton and neutron turned out to be a single particle in two forms. The two particles interconverted in the nucleus with equal ease.

A year after the discovery of the antiproton, the **anti-neutron** was uncovered: a fourth particle in one group. The antineutron didn't fit into the group scheme. The nucleon group could be viewed as made up of two pairs: proton and neutron with their antiparticles. But then the proton and the neutron would be two distinct particles. That was the **quadruplet** of nucleons.

Finally, we note that the hyperons only come in pairs. Is there any law underlying this group structure of particles? No final conclusion can yet be drawn.

Most particles and their antiparticles differ in the sign of electric charge such as the electron and positron, the proton and antiproton, the two μ-mesons and, in general, for all charged particles. But the neutron and its antineutron must have another anti-property. There is no electric charge on neutrons and their masses are the same, as in all particle-antiparticle pairs. The difference here, it appears, is in the sense of the magnetic moment. For charged particles such as electrons in atoms, the energy states team in two due to opposite spins of electrons of each energy level. Yet, the particles remain electrons, not one of them changes into a positron. The spins of atomic electrons are oppositely directed in pairs only means that the electrons themselves are moving in opposite directions. Electron spin is definitely oriented with respect to the direction of motion. For instance, if an electron is moving to the right, we may say that its spin is, for instance, directed at some angle upwards; if the movement is to the left, then downwards. As the velocity of the electron approaches that of light, the direction of its spin comes closer and closer to that of its motion. In the case of the positron, the situation is reversed. For a very fast positron, the spin is almost counter to the direction of motion.

Also, nuclear neutrons can occupy energy levels, two at a time, and no antineutron is born. **The difference in direction of the magnetic moments of a neutron and its antineutron is responsible for their anti-property. Upon encounter into field quanta, the particle and antiparticle vanish, their spins cancel each other.**

Decay Of Elementary Particles

How do particles originate and vanish? Photographic plates are the first to witness these events of the subatomic world.

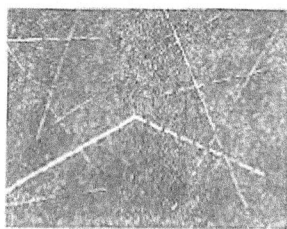

Here in the corner of the plate is a thick mu-minus- meson track. Before even reaching the middle of the plate it 'breaks' and goes off in a dashed line. This portion of the track belongs to an electron. At the breaking point, two particles were born which carried off the energy and momentum of the μ-meson that was not imparted to the electron. These two particles are a neutrino and an antineutrino. As a rule, π-mesons do not decay into electrons directly. They first generate μ-mesons. Here, too, we see that the nuclear field and the electromagnetic field are not completely separate. A particle of nuclear origin converts into a particle of electromagnetic nature.

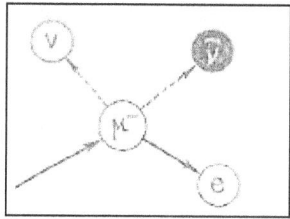

Why do π-mesons disintegrate into two particles, and μ-mesons decay into three? It's all due to the spin. **The sum of the spins of the daughter particles must equal the spin of the parent particle.** The μ-meson has half-spin, the electron also. But since the electron cannot carry off the entire mass of the μ-meson, a neutrino is needed, which takes up the residue of mass in the form of energy of motion. But the neutrino spin is also one half, then the total spin of the newly born particles is greater than the spin of the parent. The neutrino must now get rid of the extra spin. This is the **antineutrino** with opposite spin. The result is the three particles.

In the decay of a π-meson, one neutrino or antineutrino is enough, with spin counter to that of the parent μ-meson. These spins cancel, yielding zero, which is equal to the spin of the original π-meson.

In the case of hyperons, the ultimate stable product of their decay is frequently the proton. In addition, hyperons emit π-mesons.

Two worlds and two limiting types of transformations: the electron in the light particles, the proton among the heavy particles.

Two worlds and two inevitable decay satellites: the neutrino in the case of light particles, and the π-mesons in the case of the heavy particles.

The choice of the number of decay schemes appears to obey the conservation of the total charge and total spin of a particle in decay. But still these laws leave a little latitude in the choice of the decay scheme. There ought to be some other decay laws that would narrow down the pathways that unstable particles can follow for conversion into the stable building stones of matter the proton and the electron.

Studies of the processes of decay of elementary particles showed up three types that proceed with different strength and at different rates. The **strongest interactions** occur in collisions of nuclear particles in interactions in the nucleus. They are typified by large energies of the order of the energy proper of the π-meson and higher, and, accordingly (by the uncertainty relation) very short lifetimes. The time factor here is of the order of 10^{-23} second.

The next in strength and duration in the decay of elementary particles is **electromagnetic interaction**. It is in this process that electron-positron encounters produce two gamma photons. In this class too is the above-described decay of a neutral π-zero-meson into gamma photons. This process has a duration of the order of 10^{-17} second.

The **weakest and longest interaction** is responsible for the great majority of decays of elementary particles. It accounts for the decay of mu-, π- and K-mesons, the neutron and hyperons. The duration of this 'universal' decay interaction that affects particles in all groups is 10^{-10} second and more.

The K-mesons and hyperons grouped together in a different way from that of the other particles. These two groups did not fit into the classification of the other particles and were named **'strange particles'**. The special quantity that describes the degree to which they deviated from the properties is known as 'strangeness'. It was found that strange particles cannot decay into ordinary particles other than by the slow weak interaction. Strange particles are born only in pairs. The sum of strangenesses of those pairs is equal to zero, like the original, ordinary particles. In other words, in strong and electromagnetic interactions the strangeness does not change. This became known as **the law of conservation of strangeness.**

Studies of the decay of K-mesons made possible one of the biggest discoveries in the physics of elementary particles after the detection of vacuum effects that led to theory of 'nonconservation of parity'.

In particle physics, a **K-meson** (also called a **kaon** and denoted K) is any one of a group of four mesons distinguished by the fact that they carry a quantum number called strangeness. In the quark model they are understood to contain a strange quark (or antiquark), paired with an up or down antiquark (or quark).

Kaons have proved to be a copious source of information on the nature of fundamental interactions since their discovery in 1947. They were essential in establishing the foundations of the Standard Model of particle physics, such as the quark model of hadrons and the theory of quark mixing. Kaons played a distinguished role in our understanding of fundamental conservation laws: the discovery of CP violation (a phenomenon generating the observed matter-antimatter asymmetry of the universe), was made in the kaon system.

K--mesons were discovered in cosmic rays. Among the mass of bizarre tracks that cosmic particles leave on photographic plates, the vigilant eyes of physicists discerned the traces of certain new particles with masses roughly a thousand times more than the electron mass. There turned out to be three kinds of K-meson: positive, negative, and neutral. K^+, K^0, K^-, **mass:** 493.667 ± 0.013 MeV/c^2, **electric charge:** K±: ±e, K0: 0, **spin:** 0

The spin was determined and came out equal to zero. At first the family of K-mesons did not seem to differ much (with the exception of mass) from the lighter family of π-mesons: the same zero spin, the same triplet of particles, only the neutral K-mesons were heavier than their lighter cousins. The tracks left behind by K-mesons on photographic plates produced ordinary tracks that frequently terminated, giving way to thin tracks. What this meant was that the K-mesons had decayed into lighter particles. A study of the secondary tracks showed that they belonged to π-mesons.

The decay of the neutral K-meson was more difficult. Two tracks, and sometimes three, coming out of the end-point of the K-meson flight path.

Production and decay of a neutral cascade hyperon (Xi zero).

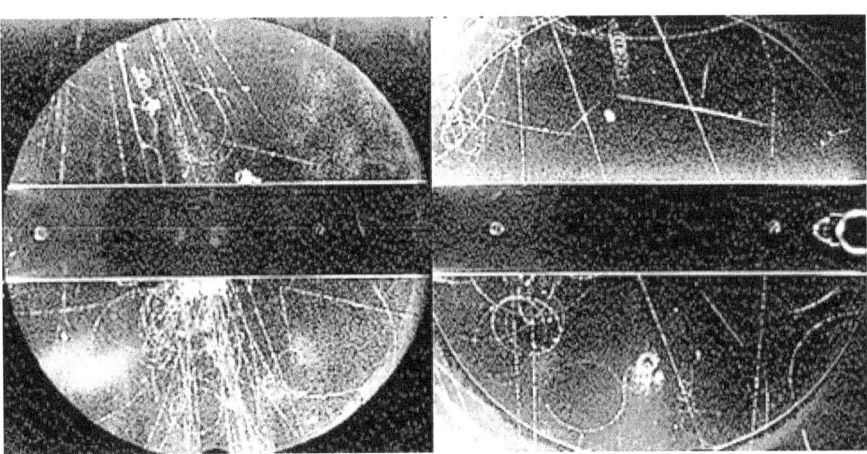

123

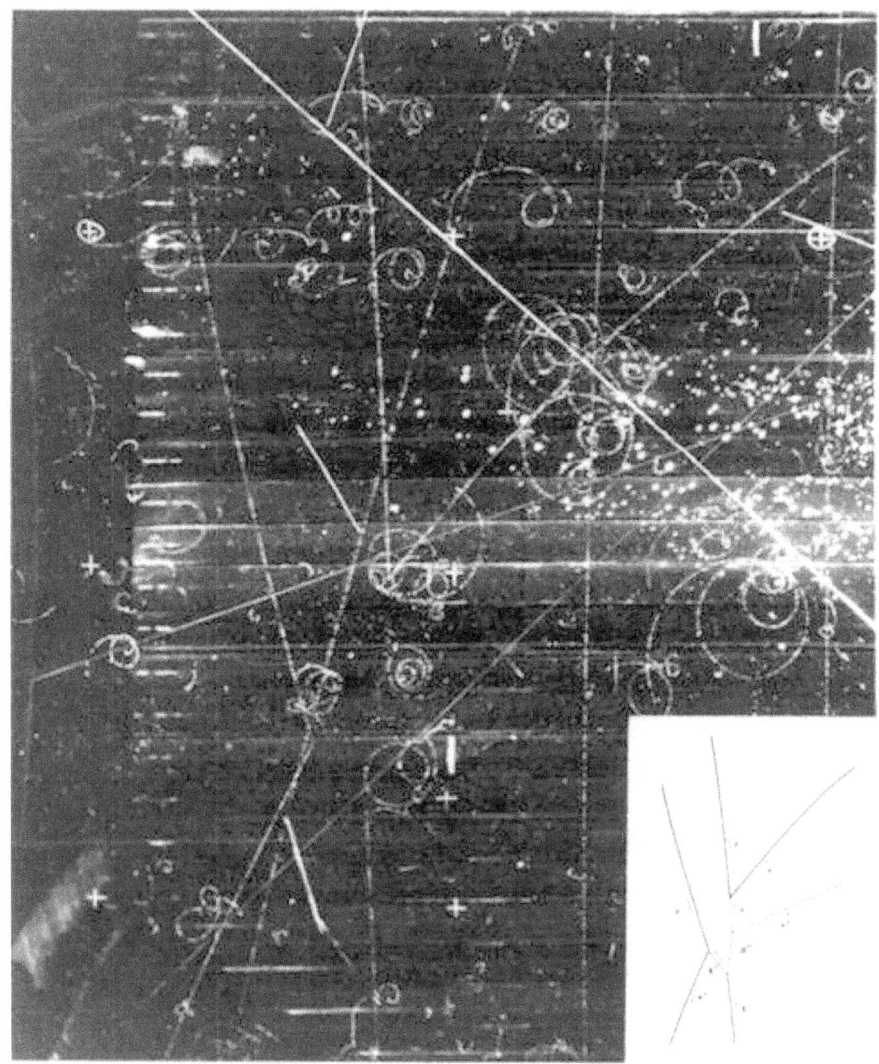

Production and decay of a neutral cascade hyperon (Xi zero)

December 20, 1947, saw the publication of the first images capturing the decay of particles we now know as kaons - the first examples of "*strange*" particles. We now know the kaons are mesons containing strange quarks. The tell-tale tracks appeared in a cloud chamber exposed to cosmic rays, which was operated by Clifford Butler and George Rochester at the University of Manchester. The left-hand image shows the decay of a neutral kaon, captured the previous year, 1946. Being uncharged, the neutral kaon leaves no track, but a "V" of tracks appears when it decays into two lighter charged particles, each of which is a *pion* (just below the central bar towards right of the chamber). The right-hand image shows the decay of a charged kaon into a *muon* and a *neutrino*. The *kaon* has come in at the top right of the chamber and the decay occurs where the track appears to bend to the left abruptly. The track beyond this kink is due to the *muon*, which penetrates the bar across the chamber. The *neutrino* has no charge and so remains invisible in the detector - **its presence is inferred from the imbalance in momentum** where the kink occurs. The decays of *kaons* are related to nuclear beta-decay, and occur through the weak force.

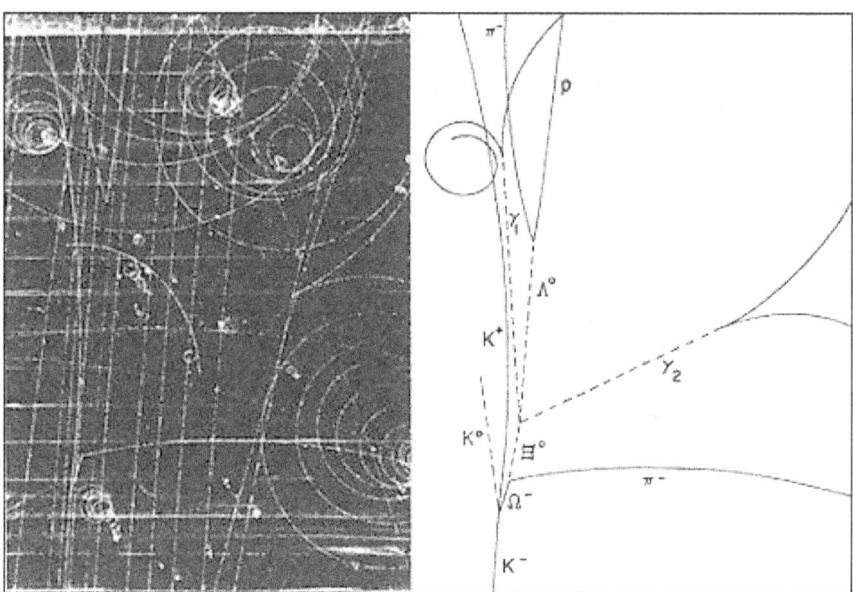

The bubble chamber picture of the first omega-minus. An incoming K- meson interacts with a proton in the liquid Hydrogen of the bubble chamber and produces an omega-minus, a K° and a K⁺ meson which all decay into other particles. Neutral particles which produce no tracks in the chamber are shown by dashed lines. The presence and properties of the neutral particles are established by analysis of the tracks of their charged decay products and application of the laws of conservation of mass and energy.

As before, all these tracks belonged to π-mesons. Thus, neutral mesons sometimes decayed into three and sometimes into two π-mesons, while all the other particles always disintegrated into the same daughter particles in one way only. This introduced two different neutral K-mesons. One of them was labeled **tau-meson**, the other the **theta-meson**: Two different mesons for the two distinct schemes of decay.

The tau-meson and the theta-meson had identical masses, but one and the same particle cannot, surely, decay first into two and then into three identical daughter particles. The law of conservation of energy does not forbid it, the conservation laws of momentum and spin have nothing against it. Yet it is forbidden by the **law of conservation of parity**, a new law established by Quantum Mechanics.

In electronic transition between energy states, the overlapping of the 'clouds of probability' of the initial and terminal states of the electron essentially related to parity. The notion of parity of the wave function is nothing but the overlapping of probabilities. From there the concept of parity was extended to the state itself as described by the wave function. Wave functions are ordinary mathematical functions, among which we frequently find sines and cosines. The sine of a negative angle is equal to the sine of a positive angle with sign reversed. The cosine will not change in the mirror. This is called "**space inversion**". To distinguish even functions from odd ones, they were given signs: plus for

even, minus for odd. An investigation of the solutions of the Schrödinger equation showed that for atomic electrons **the parity never changes in jumps to new states.** In Classical Mechanics the equivalence of directions is called the **isotropy of space**.

If the wave function of an electron was first even and then after a jump to another state became odd, this would signify only one thing- the wave function, of the photon generated in the transition is odd. Later, the concept of parity was extended from atomic states to separate particles. The photon was the first; later, labels appeared on the other particles as well. The **electron is odd particle**. The **electron spin in the mirror remains unchanged**. If it were even, the mirror image would not differ from the real thing. **The parity of the initial particle must be equal to the <u>product</u> of the parities of all decay particles produced.** So far, particles have never violated this injunction which goes by the name of the law of conservation of parity.

And now we have the neutral K-meson or keon. Judging by the fact that it decays into two π-mesons, the **keon is an even particle** (a minus and a minus produce a plus). Yet its decay into three π-mesons indicates that this particle is odd (a minus times a minus times a minus yields a minus). It is clear that we are dealing with one and not two particles: the masses of the tau- and theta-mesons coincide too closely. But then the K-meson is a particle with double parity. That violated the law of conservation of parity.

The physicists Li and Young postulated that parity can break down in the decay of K-mesons and, generally, in all weak interactions which give rise to the decay of mesons and the beta decay of nuclear neutrons. Calculations showed that if the parity did break down, then in nuclear beta decay the electrons should fly out mostly in a direction opposite that of nuclear spin. Under ordinary conditions, the nuclei orient their spins randomly and electrons come out in all directions. So the first thing that had to be done was to line up the nuclei so that all their spins would be in one direction, and then to keep them lined up during the experiment.

To keep nuclei lined up in one direction, a piece of beta-radioactive material was put in a strong magnetic field that kept the spin magnets of the nuclei aligned. Then the temperature was drastically lowered to only five hundredths of a degree above absolute zero to eliminate the distorting effects of thermal motion of the nuclei. Then a series of electron counters were arranged around this set-up at a slight angle to the direction of nuclear spin and in a mirrored direction to it. The counters were switched on and it was soon found that there were fewer electron counts in the forward direction than in the 'mirror' direction. Li and Young, and, independently, the Soviet physicist L. Landau concluded that the violation of the law of parity it was attributed to the alignment of particles.

Similarly, **the positron comprises a mirror image of an electron with reversed spin and reversed electric charge**. That is an exact reflection of the electron. When a particle is reflected by *'combined inversion'* of its spin and charge we always get its antiparticle. The neutral K-meson gives birth to the neutral, but anti- K-meson. The neutral K-mesons that were experimentally observed proved to be a mixture of two kinds: The K-zero-meson and its antiparticle.

But the K-zero-meson is odd, while its antiparticle is even. Thus, the spin of a particle could be oriented relative to the direction of motion of the particle only in some definite way, and it must be opposite to that of the antiparticle. Actually, the difference in 'spiralness', or helicity, is what distinguishes particles from their antiparticles. **Particles possess left-hand helicity and antiparticles, right-hand helicity.**

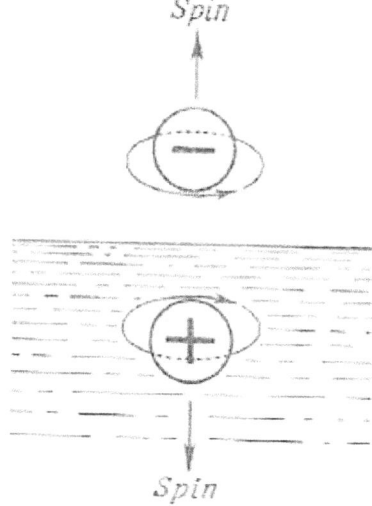

CHAPTER 6:
CRITIQUE ON QUANTUM MECHANICS

Critique On The Special Theory Of Relativity

We have seen that many contributing theories have been beaten up to extreme in order to spare Quantum Mechanics from acute demise. First was the special theory of relativity that was taken by Dirac as undisputable canon of physics. That fixed the speed of light as a universal frame of reference despite the fact that the very entity of the space void is still entirely undetermined. Secondly, Heisenberg's Uncertainty Principle was squeezed to ultimate ingredients in order to justify all violation of laws of conservation during transitions. Thirdly, de Broglie's matter waves were utilized whenever the domain of action of particles was in concern.

The special theory of relativity created a big dilemma to physicists with its audacious onslaught on Classical Mechanics. It however left physicists with greater unanswered questions regarding the new postulated constancies of the rest-mass of particles and of the reference frame of the constant speed of light. **Somehow, the special theory of relativity assumed as facts that the light picked a fixed frame in the universal space where it travels with stubborn constant speed to which we should set our clocks and weight scales.** Then, the same theory assumed as facts that particles must have constant identifiable rest-masses peculiar to their identities and without explaining why such frame of rest could be different from that of the light. It then reached the paradoxical conclusion that photons have no rest-mass but in the same time could possess indeterminate relativistic masses and act as corpuscles at very high frequencies.

Thus the growth of Quantum Mechanics was truly an intellectual development of man's effort to understand the working of the Almighty by the same mind created by such supreme Deity. That led to the conclusion that bricks of the atom could not be fixed, measured, or visualized. That was attributed to the same distilled theories of the wave properties that smear out the particle in space. These wave properties are an external manifestation of the interrelations of particles with their fields. In other words, an electron is clouded out due to its interaction with other particles, including electrons.

An electron in virtual fashion emits photons and interacts with photons emitted by other particles. The result is either mutual repulsion or attraction of the particles. The electron is, as it were, wrapped up in a cloud of virtual photons which it is emitting and absorbing. This cloud is boundless. There will always be low-energy photons such that the Heisenberg relation will allow them to go to any distance from the electron that emitted them. It is the photon cloud which smears the electron out in space that does not allow us to speak of *exact dimensions*.

Still, the cloud of photons rapidly contracts as it approaches the core of the particle. At distances of the order of 10^{-11} cm, where virtual photons have energy enough to form electron-positron pairs, we have what might be called a '*trembling*' electron. The electron is still clouded out, yet over a smaller region of space. Again there are no exact dimension. Maybe it would be possible to measure exactly the size of a 'bare' electron, rid of its photon and electron-positron clouds. But, both de Broglie's matter wave and Heisenberg's uncertainty principle render such bare electron impossible, for **there is no such thing in nature as a noninteracting electron. The particle and its interaction are two things that form a unity**, an inseparable unity.

The only thing left for us to suppose is that inside all these clouds is something in the nature of a '*core*', as physicists have termed it. But today we know nothing definite about what this core looks like and of what goes on inside it. Physicists have tried to do a similar job in describing the structure of another fundamental particle, the proton. The proton emits π-mesons with energies that are naturally not less than their rest energies. π-mesons have very short lifetimes. They cannot travel too far from the proton that produced them. Indeed, the dimensions of the π-meson cloud about the proton are very small, of the order of 10^{-13} cm. Unlike the electron, the proton is only very slightly clouded out by π-mesons. But we know that protons are rather energetic in their interactions with K-mesons as well. A proton can also have a virtual cloud corresponding to this interaction, a K-meson cloud.

Since the rest-energy of a K-meson is some three times that of the π-meson, the K-meson cloud should be just that number of times smaller and should be located inside the π-meson cloud. Still deeper in the interior we should find the concentrated '**trembling proton**' that decays virtually into proton-antiproton pairs.

Thus, the structure of particles of the subatomic world is a reflection of all their interactions with other particles. Particles do not exist without their interactions. All elementary particles are interrelated by interactions. Interaction is an integral and natural part of their structure. Conversely, the character and degree of interaction are determined by the structure of the particle.

The Schrödinger equation was constructed along the lines of the classical wave equation with the sole difference that it described not ordinary waves but 'waves of probability', which expressed the law of motion of elementary particles in space and time.

At first the satisfaction was complete. The elementary particles willingly obeyed these laws. True, from the very start of Quantum Mechanics it was found that these old concepts would not function well in the new physics.

The Heisenberg's uncertainty relation made it evident that the earlier ideas of exact position and velocity, particle energy and time could be applied in the subatomic world on a very restricted scale. Soon, the elementary particles gained energy sufficient for mutual transformations and cracks appeared in the old Schrödinger's wave equation. The semi-classical method for establishing the laws of motion of particles in space and time broke down completely. The wave function was not in a position to describe conversion.

The wave equation treated conversion of particles and waves according to the familiar laws of conservation of energy and momentum. In this approach, **the process of transformation of matter and wave itself was left out of consideration.** At the instant of transformation, the particle is not in motion in the ordinary sense of the word. Yet, the wave equation made no distinction of such singularity or discontinuity of the state of the particle. Secondly, because one type of particle vanishes and a different type of particle appears, the equations of motion referred always to a single invariable type of particle. Which means that this classical approach to phenomena of the subatomic world carried over into Quantum Mechanics by means of the space and time concepts was clearly insufficient. That did not reflect the basic essence of this world, the transformations of particles into one another and into field quanta, and also the reverse conversion of quanta into particles of matter.

The problem now was to determine the actual course of transformation. But this required a radical change in the mode of description. Quantum Mechanics did this by introducing those virtual processes when uncertainty ensues during changes in coordinates, time, and energy. Those virtual processes too fail and do not yield a final solution to the problem. A still more profound approach is needed in which the classical conceptions of space and time will probably undergo fundamental change.

The limit of these techniques is, for length, of the order of the range of nuclear forces, or 10^{-13} centimeter, and for time of the order of the 'nuclear time', i.e., 10^{-23} second. Some scientists believe that the 'length quantum', if it exists, should be hundreds or even thousands of times shorter. It is understandable why we never notice the existence of quanta of space and time. They are simply too small. And the same goes for lengths.

Indeed, **what is mass? What is the mass of a proton? What is matter?**
When we say that the proton has a mass of approximately 10^{-24} gram, we only mean that one gram of substance contains roughly 10^{24} protons. Thus, to define mass as a measure of substance for protons and other elementary particles is rather meaningless. The second definition of mass is that of the inertia of a body or the resistance the body offers to any change in its state. Here, mass presents the *interaction with a field*. That is exemplified with the mass pumped into an electron when accelerated. The accelerated electron acquires extra mass from the field; when its speed diminishes, it returns this mass to the field. Small as these portions of acquired and lost mass **may be** they do exist. Hence, mass is a variable quantity and thus loses its property of a definite measure.

Thus, Classical Mechanics contradicts its underlying axioms. It started by implying that mass was inertia, which requires interaction with mobile field. But it required that such mobile interaction to be constant. Thus, according to Classical Mechanics, mass must be constant and mobile in the same time. That spells failure. If Classical Mechanics failed to describe the mass content of matter, the laws of conservation of energy and momentum must fail as well, since they are tied directly to mass. That was not a surprise. As soon as the electric charge was discovered, Classical Mechanics offered no account for the negative and positive **charges**. That was followed by the **spin, parity, strangeness**, none of which has a relative in the classical world.

Thus we find in the subatomic world that mass itself has to be measured with something that cannot be measured. That is the field. Then the special relativity borrowed the classical mass and related to the proper rest-mass of a particle in terms of the speed of light. Doesn't it then follow that the rest-mass is also a measure of inertia? We recall that when the kinetic energy of particles is compared with their energy proper as determined precisely by the rest-mass, particles obtained the possibility of actual transformations into the quanta of its field. Then the **rest-mass becomes a measure of the qualitative stability of particles**. For some particles this mass is not very great and conversion into quanta can begin at rather low energies. In the case of other particles, it is much greater, and accordingly the particles are considerably more stable. All this makes mass a very intricate concept. On the one hand, mass is some kind of characteristic of the particle as such; on the other, mass is a determining factor in all interactions of the particle.

Today, all the problems involved in determining the deep inner essence of entities of the subatomic world come up against this greatest of unconquered peaks of physics: The interrelation of the two basic forms of matter substance and the field. Particles of substance possess properties of the field. Field quanta have material properties. In the years 1860's, when physics had just acquired the concept of the field, matter was perceived as the initiator of fields. The particles of a substance generate a field about themselves. The field is only an auxiliary tool for handling particle interactions. There is no field without matter. But time passed, and it was found that **a field could generate particles**, that **particles could vanish and become a field** not as auxiliary as might be supposed.

Do our notions about the existence and interconnection field and matter possess any degree of truth?
Science develops in such a way that new conceptions originate very slowly. After all, human beings live in the world of ordinary things, common notions, and their minds hold tenaciously to these notions. It is very difficult to make the translation to the unconceivable conceptions that make for a correct picture of inaccessible worlds. Quantum Mechanics was able to combine the old concepts into new **particle-wave**, **positron-hole** and **meson-quantum** images. But in the minds of physicists these dual entities have not fully merged into unified actuality. This merging might never occur but the underlying causes for such illusive truth might lead us to greater but different truth.

Critique On The Theory Of Matter Waves
The great merit of the new theory of matter waves lies in the fact that it has solved great number of problems connected with the electronic structure in the atom and spectral lines and the nuclear reaction. The theory leads to equations whose solutions correspond to definite energy values, comparable to the energy terms of Bohr's atom model. Thus the quantum numbers of atomic structure and spectroscopy are inherent in the theory and are given in fundamental terms, not as arbitrary assumptions.

The very discreteness of atomic processes appears as a natural outcome of the theory. Wave Mechanics also accounts for facts which are inexplicable by the older theories, such as half-integer quantum numbers, zero-point energy, etc. It has even led to predictions which later have been verified experimentally, such as diffraction of electrons, resonance phenomena in spectroscopy, etc.

Furthermore, events impossible to the classical theory are found, when treated by wave mechanics, to have a very small but finite probability of taking place. On the basis of this prediction, natural radioactive process and artificial disintegration have received a very satisfactory explanation. Application of Wave Mechanics to these specifically nuclear phenomena has resulted in some understanding of the nucleus itself including its structure and energy content.

Further, the general behavior of matter in the atomic state has been interpreted to a high degree of satisfaction by Wave Mechanics. It has shown that not only radiation but also matter exhibits sometimes wave-like properties and sometimes corpuscular properties; it has further led the way to treat the two concepts, wave and corpuscle, as complementary aspects of viewing one and the same objective process, which only in definite limiting cases admits of complete pictorial representation and is therefore nearly always subject to certain indefiniteness, given by the principle of uncertainty.

(i) Herein lies the great advance made by the new mechanics over the absolute determinism of classical mechanics. In general, if **the doctrine of determinism** were to be rejected, there would be no order or regularity in physical phenomena and no scientific knowledge of them could exist. We know beyond any doubt that most elementary particles possess very determined identities attached to stringiest mathematical constraints. We also have no doubt that the speed of light is a universally determined constant. Thus, determinism is an actual fact and has proved its value by its progress and the number of its practical applications. We have a convincing proof of this in the modern art of artificial transmutation of elements, which is possible only if there exists a certain amount of determinism coupled with indeterminacy in the processes of interaction of material bodies.

(ii) **Planck's constant 'h'** which is considered to be the limiting barrier of determinism and a measure of indeterminacy, is really an enigma of Modern Physics..

When one considers the way in which the constant h enters into Atomic Physics, viz., through the corpuscular aspect in the case of radiation, but through the wave aspect in the case of matter, it becomes possible to surmise that double complementary aspect of ***determinism of corpuscles and indeterminacy of field***. For, on this assumption alone one could understand why an obviously wave form radiation acts as corpuscles and corpuscular matter behaves like a wave.

(iii) **The very concept of probability,** which forms the basis of the principle of indeterminacy, implies the complementary double aspect of determinism and indeterminacy in every physical phenomenon. When the history of a single physical unit, whether electron or photon, is investigated, the associated wave symbolizes the probabilities of localization and the dynamic state of that unit; but when we are dealing with a very great number of identical units, for instance, a beam of electrons, then the wave represents the statistical distribution of the totality of units. Hence, in both cases statistical law must be used. Now a statistical law is one which defines the more or less constant mode of action of a great collection of similar things without stating anything about a particular individual of the collection, except in so far as it, belongs to the collection. A statistical law is a ***law of averages*** and there is no finality or exactness about it when applied to particular individual cases. The reason for this is that when forming part of a system, a physical unit loses a large measure of its individuality, the latter tending to merge in the greater individuality of the system. To make **a** real individual of a physical unit belonging to a system, then, it is necessary to take this unit from out of the system, *i.e.,* to break the link which binds it to the system. The physical unit cannot be observed perfectly so long as it forms part of the system and the system is impaired once the individual unit has been identified. The concept of the physical unit thus becomes completely clear and properly defined, only if it is regarded as a unit completely independent of everything else. Applying this general idea to a single particle we have a system formed of localization and dynamic state of the particle. The two lose their individualities to a large measure when they form part of the system. Hence, ***when we try to define exactly one of them, the other becomes most ill-defined.*** The same consideration can be applied to a system made up of a very great number of particles. Thus a close analysis of physical reality, as an individual unit or as a system of units, brings out clearly the intimate and unavoidable presence of the two aspects of determinism and indeterminacy in every physical phenomenon.

CHAPTER 7:
CONTRIBUTION BY NATIONS OF SCIENTISTS

Italy:
Leonardo Da Vinci (April 15, 1452 -May 2, 1519), Italy.
Guglielmo Marconi (25 April 1874- 20 July 1937), Italy
Alessandro Giuseppe Antonio Anastasio Volta (February 18, 1745- March 5, 1827), Italy,
Galileo Galilei (February 15 1564- January 8 1642), Italy.
Leonardo Da Vinci (April 15, 1452 -May 2, 1519), Italy.
Cecil Frank Powell (5 December 1903- 9 August 1969), Italy.
Giuseppe Occhialini (December 5, 1907 - December 30, 1993), Italy.

England:
James Clerk Maxwell (13 June 1831-5 November 1879), Scotland.
Sir William Crookes (17th of June 1832 -4th of April 1919), England.
Michael Faraday (22 September 1791 – 25 August 1867), England.
William Prout FRS (15 January 1785 – 9 April 1850), England.
William Prout (1785-1850), England.
John Dalton (September 6th 1766 - July 27th 1844), England.
Thomas Young (1773-1829), England
James Watt(19 January 1736 – 25 August 1819), Scotland.
Isaac Newton (January 4 1643 – March 31 1727), England.
Sir Joseph John "J. J." Thomson (18 December 1856 – 30 August 1940), England.
James Chadwick (October 20, 1891-July 24, 1974), England,
Paul Dirac (8 August 1902- 20 October 1984), England,
Ernest Rutherford (August 30, 1871- October 1937), New Zealand,
William Lawrence Bragg (31 March 1890- July 1971), England.
Patrick Maynard Stuart Blackett (18th November, 1897- 13 July 1974), England

Germany:
Georg Simon Ohm (16 March 1789 – 6 July 1854), Germany.
Walther Ludwig Julius Kossel (4 January 1888-- 22 May 1956), Germany,
Arnold Johannes Wilhelm Sommerfeld (5 December 1868 – 26 April 1951), Germany,
Pauli, Wolfgang (1900-1958)
Max Born (11 December 1882-5 January 1970)
Werner Heisenberg (5 December 1901-1 February 1976), Germany,
Johannes Stark (1874-1957), Germany
Joseph von Fraunhofer (6 March 1787 – 7 June 1826), Germany.
Albert Einstein (14 March 1879 – 18 April 1955), Germany.
Max Planck (April 23, 1858 – October 4, 1947), Germany.
Wilhelm Röntgen (27 March 1845- 10 February 1923), Germany.

France:
André-Marie Ampère (1775-1836), France.
Augustin-Jean Fresnel (10 May 1788 - 14 July 1827), France.
Charles-Augstin de Coulomb (June 14th 1736-August 23 1806), France.
Joseph-Louis Lagrange (25 January 1736 - 10 April 1813), France.
Blaise Pascal (June 19, 1623 - August 19, 1662), France.
Marie Curie (November 7, 1867- July 9, 1934), France.
Antoine Henri Becquerel (15 December 1852 – 25 August 1908), France.
Louis de Broglie (15 August 1892 – 19 March 1987), France,

Denmark:
Hans Christian Ørsted (August 14 1777 - March 9 1851), Denmark.

Belgium:
Simon Stevin (1548-1620) Belgium.

Austria:
Samuel Goudsmit [1902-1978], Ludwig Eduard Boltzmann (February 20, 1844 – September 5, 1906), Austria.
Erwin Schrödinger (August 12, 1887- January 4, 1961), Austria

USA:
Robert A. Millikan (March 22, 1868- December 19, 1953), USA,
George E. Uhlenbeck 1900, Netherlands -1988, US

Denmark:
Niels Henrik David Bohr (October 7, 1885 - November 18, 1962), Denmark,

Russia:
Yakov Il'ich Frenkel (10 February 1894- 23 January 1952), Russia.
N. N. Bogolubov (Aug. 21, 1909- Feb. 13, 1992), Russia,
Igor Tamm (8 July 1895-12 April 1971), Russia.
Nikolai Lobachevsky (1792 -1856), Russia

Japan:
Hideki Yukawa (23 January 1907-8 September 1981), Japan.

CHAPTER 8:
HIGHLIGHTS OF MAIN FACTS

1. The Ancient Greek believed that the entire universe consisted of four elements *water, air, earth, and fire*.

2. The supreme creator was viewed as being able to create live matter from nothing. Hence came the theory of '*creation from nothing*'. The supreme deity of the past fictional world is still supreme in our sensual world.

3. Galileo was even forced to proclaim that his statements of the **Earth** not being the center of the Universe, where only 'vague assumptions' in stead of facts.

4. The genius of Newton lies in the simplicity of the three laws that offered a measure for the **invariability of motion** in terms of the *total energy* and *angular momentum* of the moving particles.

5. Newton's laws had been honed in harnessing the steam energy which started a new era of *powerful engines*.

6. Charles-Augstin de Coulomb discovered **Coulomb's law** describing the electrostatic interaction between electrically charged particles, first published in 1783. The magnitude of the electrostatic force between two point electric charges is directly proportional to the product of the magnitudes of each of the charges and inversely proportional to the square of the distance between the two charges.

7. John Dalton formulated the modern atomic theory, discovered information on gas laws, atomic weight. Dalton's atomic theory held: *The atoms of a given element are different from those of any other element; the atoms of different elements can be distinguished from one another by their respective relative atomic weights. All atoms of a given element are identical.* Atoms of one element can combine with atoms of other elements to form chemical compounds; a given compound always has the same relative numbers of types of atoms. Atoms cannot be created, divided into smaller particles, nor destroyed in the chemical process; a chemical reaction simply changes the way atoms are grouped together. Elements are made of tiny particles called **atoms**.

8. William Prout observed that the atomic weights that had been measured for the elements known at that time appeared to be *whole multiples of the atomic weight of Hydrogen*. He then hypothesized that the **Hydrogen atom** was the only truly fundamental object and that the atoms of other elements were actually groupings of various numbers of Hydrogen atoms.

9. Alessandro Volta introduced the first **battery** in 1800, made from alternating layers of zinc and copper, which provided scientists with a more reliable source of electrical energy than the electrostatic machines previously used. Volta described the relationships between electrical capacitance (C), electrical potential (V) and charge (Q), and formulated Volta's Law of capacitance.

10. Hans Christian Ørsted proved that an electric current produces a circular **magnetic field** as it flows through a wire.

11. Michael Faraday discovered **electromagnetic induction**, diamagnetism, and laws of electrolysis. He established that magnetism could affect rays of light and that there was an underlying relationship between the two phenomena.

12. André-Marie Ampère formulated the **Ampere's Law** it relates the integrated magnetic field around a closed loop to the electric current passing through the loop.

13. Georg Simon Ohm determined the direct proportionality between the potential difference (voltage) applied across a conductor and the resultant electric current, Hence, **Ohm's Law**.

14. Every chemical element has its own **characteristic spectrum**.

15. In 1898, Thomson believed that the electrons emerged from the atoms of the trace gas inside his cathode ray tubes. He thus concluded that atoms were divisible, and that the **electrons** were their building blocks. Thomson described atoms as clouds of positive charge within which floated negative electrons in quantities sufficient to balance the charge.

16. Röntgen showed that the **X-rays** are produced by the impact of cathode rays on a material object.

17. No single canon in Classical Mechanics for the atom was able to account for the mysterious release of energy by Uranium, Radium and other chemical elements-a radiation of energy that continues without interruption for many thousands and millions of years without any outside source. **Radioactivity** was discovered in 1895.

18. From the Rayleigh-Jeans law the intensity of **radiation from a heated body** was supposed to increase without bound.

19. Planck concluded that the energy of radiation (like matter itself) is atomistic and that it is released and acquired not continuously but in small portions, **quanta**, as Planck called them, from the Latin 'quantum' meaning quantity.

20. Planck postulated that **quanta differ for different types of radiation**. The shorter the wavelength of light, that is the higher its frequency (in other words, the 'more violet' it is), the larger the quanta.

21. For each metal studied, there appeared to be a certain **limiting wavelength** of incident light. When this wavelength was exceeded, the electrons in the flask disappeared at once and the current ceased to flow no matter how strong the light was.

22. Einstein postulated that light is simply a **stream of quanta of energy**, all the quanta of a single wavelength being exactly the same, which is to say that the quanta carry identical portions of energy. Later, these quanta of light energy were given the name photon.

23. If the wavelength of the light was too large, the **photons** do not have energy enough to dislodge electrons from the metal.

24. In 1924, Louis de Broglie postulated the possible existence of **matter waves**. de Broglie maintained that these waves are generated in the motion of any body, whether a planet, a stone, a particle of dust or an electron.

25. As soon as the **wavelengths of photons** become small enough, they begin to act like particles and are able to knock electrons out of a metal. The best example is gamma rays, the shortest of all known electromagnetic waves.

26. Bohr estimated the radius, energy, and frequencies of his **privileged electronic orbits** with total disregard to the temporal dynamism of neighboring electrons on each individual orbit. Even the nucleus of the atom must bounce or orbit as the rest of electrons impart attraction and repulsion while in motion. Indeed, Bohr Atom Model froze in time and place in utter detachment from the balances of forces in the atom.

27. The entire positive charge of the atom appeared to be concentrated in the tiny central part, the **nucleus**.

28. Bohr rejected the concept that an electron in an atom need not give of light even when in accelerated motion and postulated the existence of **privileged orbits** about the nucleus where electrons move in steady state energy levels.

29. **Emission and absorption of radiation** occurs when electrons jump from one orbit to another.

30. The permitted circular orbits of the electron are those in which the **angular momenta** are integral multiples of $h/2\pi$.

31. Even if a **medium** was found or postulated, we still have to find out the constituents of such medium. As such, ether is as good as a field or a wave as far as the inevitability of comprehending the very essence of matter and wave. Einstein's bent space geodesics offers nothing new other than that matter bends the space, which is, by itself, undetermined.

32. There are eight possible types of **chemical behavior** of atoms in accord with the number of electrons in their outer shell.

33. Atoms are divided into *givers* and *takers*. Those with less than four electrons in their outer shell are givers. Those with more than four are takers. Naturally, it is easier to acquire two electrons, say, than to give up six such as in the situation of the Oxygen atom.

34. On these molecular energy levels, electrons still obey the **Pauli's Exclusion Principle**. Only two electrons are allowed.

35. When an atom emits **beta rays**, it does not become ionized, it does not acquire an electric charge. This strengthened the idea that the beta and gamma radiation originates in the atomic nucleus.

36. The charge of the nucleus must be equal to the **collective charge** of all the electrons in the outer structure, but with opposite sign (positive). Otherwise the atom would not be neutral the way it is.

37. Their total spin must be equal to the **sum of the spins** of the constituent particles.

38. There are serious doubts about **nuclei** consisting of protons and electrons.

39. Variation in the neutron architecture of the nucleus led to isotopes-variations of a single element. The nucleus of Tin, for instance, has ten stable isotopes.

40. Yukawa postulated the existence of a new **exchange particle** with a positive or a negative charge equal to the magnitude of the proton charge (or electron charge) and a mass approximately 200 to 300 times greater than the electron mass.

41. The nuclear forces are now attributed a **meson exchange** between protons and neutrons.

42. **Helium nucleus** is the most stable nucleus in nature.

43. There is something like a system of **shells in the nucleus**.

44. The nucleus of Uranium-235 is fissionable by a neutron of only a very **definite energy**.

45. The fact that the relativistic mass, $m(v)$, could attain infinity is in total conflict with Planck's hypothesis that **nothing in physics is boundless or without limit**.

46. **Rest-energy** of the body has no place in Classical Mechanics. It is something very special to Quantum Mechanics. We do not even know whether the concept of rest-mass is without flaws. For why do material objects not move at uniform speed of light from the instant of the beginning of the universe? or why light

choose to have a fixed frame of reference other than ours and to which we have to compare our time, mass, and length?

47. Dirac postulated that the **void vacuum**, which does not contain a single particle except the sole electron, is not empty at all. Quite the contrary, it is filled to overflowing with electrons.

48. No instruments can detect the **vacuum electrons** until they jump out of their well.

49. Dirac accepted [as facts] that particles must have very identifiable **constant rest-masses** and that light must travel at very identifiable and constant speed. Further, Dirac reached the conclusion that energy was the supreme creator of particles from void.

50. Why are particles and light divested with such very identifiable **constancies of rest-mass or speed**?
Why is energy alone divested with the **authority to create elementary particles** from void?

51. The **electron and hole** originate out of 'nonexistence' only in pairs.

52. Space that is absolutely homogeneous in the absence of bodies loses this **homogeneity** when a body is 'introduced' into it.

53. **Space** has become a repository not only of bodies, but also of fields.

54. Particles can never move with the **velocity of propagation** of the field, and the field can never propagate with a different velocity.

55. It was found that 'true' **particles of matter can have only a spin equal to one-half the modified Planck constant $h/2\pi$,** whereas field quanta must have spin equal to zero or to an integral number of $h/2\pi$. It was found that the magnitude of spin exerts an essential influence on the behavior of microentities which was postulated by the Pauli's Exclusion Principle. It requires

56. **All π-mesons** have zero spin and hence can serve as field quanta except that π-mesons have nonzero rest-mass

57. **Neutrons** are not composites of protons and π-mesons.

58. The π-meson is certainly the biggest **hybrid of particle and quantum** yet.

59. **Charges of the new elementary particles** can have three values: +1, 0 and -1, where -1 is the charge of the electron. The spins have three values as well: 1, 1/2, and 0 in Planck units $h/2\pi$.

60. Charged π-**mesons** are more massive than the neutral particle, which naturally should be of nuclear origin entirely.

61. The difference in direction of the **magnetic moments** of a neutron and its antineutron is responsible for their anti-property. Upon encounter into field quanta, the particle and antiparticle vanish, their spins cancel each other.

62. The **sum of the spins** of the daughter particles must equal the spin of the parent particle.

63. The choice of the number of **decay schemes** appears to obey the conservation of the total charge and total spin of a particle in decay. But still these laws leave a little latitude in the choice of the decay scheme. There ought to be some other

64. An investigation of the solutions of the Schrödinger equation showed that for atomic electrons the **parity never changes in jumps to new states**. In Classical Mechanics the equivalence of directions is called the isotropy of space.

65. The electron is odd particle. The **electron spin** in the mirror remains unchanged. If it were even, the mirror image would not differ from the real thing.

66. The parity of the initial particle must be equal to **the <u>product</u> of the parities** of all decay particles produced. So far, particles have never violated this injunction which goes by the name of the law of conservation of parity.

67. The **positron** comprises a mirror image of an electron with reversed spin and reversed electric charge.

68. Particles possess left-hand helicity and antiparticles, right-hand **helicity**. The law of **parity**.

69. The **rest-mass** becomes a measure of the qualitative stability of particles. For some particles this mass is not very great and conversion into quanta can begin at rather low energies.

REFERENCES

A Course of theoretical physics, I, by A. S. Kompaneyets (1978)

ABC's of Quantum Mechanics, by V. Rydnik (1966)

Atomic Physics, by Max Born (1969)

Atomic Physics, by J. B. Rajam (1966)

Introduction to Quantum Mechanics, by Linus Pauling and E. Bright Wilson (1963)

Quantum Mechanics, by Leonard I. Schiff (1968)

Max Planck, Survey of Physics, 1925.

F. H. Newman. Recent Advances in Physics, 1934.

E. N. da C. - Andrade, The Structure of the Atom, 1934.

A. E. Ruark and H. C. Urey, Atoms, Molecules and Quanta, 1930.

G. Castelfranchi, Recent Advances in Atomic Physics, (2 Vole.), 1932.

A. Hass, Theoretical Physics, (2 Vols.), 1929.

Harnwell and Livingood, Experimental Atomic Physics, 1933.

F. K. Richtmyer, Introduction to Modern Physics, 1934.

H. A. Wilson, Modern Physics, 1944.

Max Born, Atomic Physics, 1954.

G. F. M. Jauneey, Modern Physics, 1946.

G. Joos, Theoretical Physics, 1950.

J. D. Stranathan, The Particles of Modern Physics, 1942.

R. A. Milliken, Electron (+ and -), Protons, Photons, Neutrons, Mesotrons and Cosmic Rays, 1946.

E. Grimsehl, Physics of the Atom, 1949.

H. Semat, Introduction to Atomic Physics, 1947.

C. T. Chase, The Evolution of Modern Physics, 1947.

S. Tolansky, introduction. to Atomic Physics, 1948.

F. K. Richtmyer and E. H. Kennard, Introduction to Modern Physics, 1946.

F. K. Richtmyer, E. H. Kennard and T. Lauritsen, Introduction to Modern Physics, 1956.

Oldenberg, Introduction to Atomic Physics, 1949.

W. Finklenburg, Atomic Physics, 1950.

S. Dushman, Fundamentals of Atomic Physics, 1951.

F. W. Van Name, Modern Physics, 1952.

C. Kittol, Introduction to Solid State Physics, 1954.

R. S. Shankland, Atomic and Nuclear Physics, 1955.

J. S. Townsend, Electricity do Gases, 1914.

J. J. Thomson and G. P. Thomson, Conduction of Electricity through Oases,
Vol. I (1928). Vol, II (1932)..

K. G. Emedeus, Conduction of Electricity through Oases, 1928.

K. K. Darrow, Electrical Phenomena in Oases, 1932.

A. AL Tyndall, Mobility of Positive Ions in Gases, 1938.

R. A. Millikan, The Electron, 1926.

0. W. Richardson, The Electron Theory of Matter, 1916,

D. Grimes, Meet the Electron, 1944.

E. C. Stoner, Magnetism, 1930.

J. H. Van Vleck, The Theory of Electric and Magnetic Susceptibilities, 1932.

F. Bitter, Introduction, to Ferromagnetism, 1937.

O. W. Richardson, The Emission of Electricity from Hot Bodies, 1921.

A. L. Riemann, Thromionic Emission, 1934.

H. S. Allen, Photoelectricity, 1925.

A. L. Hughes and L. A. du Bridge, Photoelectric Phenomena, 1932.

K. H. Spring, Photons and Electrons, 1950.

J. Stokley, Electrons in Action, 1946,

L. R. Koller, Physics of Electron Tubes, 1937.

F. E. Terman, Fundamentals of Radio, 1938.
W. L. Everitt, Fundamentals of Radio and Electronics, 1958
A. L. Albert, Fundamental Electronics and Vacuum Tubes, 19-18.
L. B. Arguimban, Vacuum tube circuits and transistors, 1956
Admiralty, Handbook. of Wireless Telegraphy, (2 Vols.) 1939.
A. A. Ghirardi, Radio Physics Course, 1933.
J. H. Rayner, Modern Radio Communication, 1947
W. F. Lovering, Radio Communication, 1958
J. L. Hornung, Radar Primer, 1948.
Radar School, Principle of Radar, 1946.
E. C. Pollard and J. M. Sturtevant, Microwaves and Radar Electronics, 1918.
W. Gordy, W. V. Smith & R. F. Trambaru o, Microwave Spectroscopy, 1953.
AL W. P. Strand berg. Microwave Spectroscopy, 1954.
H. Motz, Electromagnetic problems of Microwave theory, 1951.
Campbell and Ritchie, Photocells, 1934.
Zworykin and Wilson, Photocells, 1934.
-0. A. Briggs, Sound Reproduction, 1950.
G. F. Jones, Sound Film Reproduction, 1936.
V. K. Zworkyin and E. G. Ramberg, Photoelectricity and its applications, 1949.
H. K. Henisch, Metal Rectifiers, 1949.
J. T. Mac Gregor Morris and J. A. Henly, Cathode Ray Oscillography, 1936.
J. H. Rayner, Cathode Ray Oscillography, 1945.
W. C. Eddy, Television The Eyes of Tomorrow), 1945.
J. H Boyner, Television, 1934.
M. G. Scroggie, Television, 1935.
A. Dinsdale, First Principles of Television, 1932.
G. V. Dowding, Practical Television, 1935.
E. F. Burton and W. H. Kohl, The Electron Microscope, 1946.

F. W. Aston, Mass Spectra and Isotopes, 1933,
J. J. Thomson, Rays of Positive Electricity, 1923.
Reviews of Modern Physics, Nuclear Physics C., July 1937.
A. H. Compton and S. K. Allison, X-Rays in Theory and Experiment, 1935.
Maurice de Broglie, X-Rays, 1925.
0. W. G. Kaye, X-Rays, 1926.
W. H. Bragg and W. L. Bragg, X-Rays and Crystal Structure. 1918.
M. J. Buerger, X-Ray Crystallography, 1949.
B. L. Worsnop, X-Rays, 1930.
W. H. Zachariassen, Theory of X-Ray Diffraction in Crystals, 1915.
R. W. James, X-Ray Crystallography, 1930.
K. Lonsdale, Crystals and X-Rays, 1948.
M. Siegbahn, Spectroscopy of X-Rays, 1925.
A. J. C. Wilson, X-Ray Optics, 1949.
G. L. Clark, Applied X-Rays, 1940.
E. Rutherford, Radioactivity, 1905,
W. H. Bragg, Studies in Radioactivity, 1912.
E. Rutherford, J. Chadwick and C. D. Ellis, Radiations from Radioactive Substance, 1930
G, Newsy and F. A. Paneth, A Manual of Radioactivity, 1938.
K. K. Darrow, Bell Telephone, Some Contemporary Advances in Physics-XII
Radioactivity, 1927.
Madame Pierre Curie, Radioactivity, 1935.
A. Einstein, Relativity : Special and General Theory, 1922.
L, Silberstein, The Theory of Relativity, 1914.
H. Schmidt, Relativity and the Universe, 1921.

A. Hass, Introduction to Theoretical Physics. (Vol. II), 1929.

Sir James Jean?, Through Space and Time, 1934.

W. H. M. C. Crea, Relativity Physics, 1935.

A. S. Eddington, Space, Time and Gravitation, 1921,

E. Cunningham, The Principle of Relativit , 1921.

Max Born, Einstein's Theory of Relativity, 1922.

H. Dingle, The Special Theory of Relativity, 1940.

G. J. Whitrow, The Structure of the Universe, 1949.

G. Y. Rainich, Mathematics of Relativity, 1950.

A. Einstein, The Meaning of Relativity, 1950.

L Barnett, The Universe and Dr. Einstein, 1952.

F. Reiche, The Quantum Theory, 1930.

W. Heisenberg, The Physical Principles of the Quantum Theory, 1930.

L. Infeld, The World in Modern Science, Matter and Quanta, 1934.

W. Heitler, Quantum Theory of Radiation, 1936.

G. Temple, An Introduction to Quantum Theory, 1931.

Louis de Broglie, Matter and Light, 1939.

J. Jeans, The New Background of Science, 1947,

G. Birtwistle, New Quantum Mechanics, 1928.

D. Bohm, Quantum Theory, 1952.

J. Frenkel, Wave Mechanics, 1934.

E. Schrödinger, Four Lectures on Nave Mechanics, 1929.

N. F. Mott, Wave Mechanics, 1930.

A. Sommerfeld, Wove Mechanics, 1930.

Louis de Broglie, An Introduction to the Study of Wave Mechanics, 1930.

H T. Flint. W woe Mechanics, 1931.

F. G. Kemble, The Fundamental Principles or Quantum, Mechanics, 1937.

P. A. M. Dirac, The principles of Quantum Mechanics, 1947.

W. Wilson, Relativity and Quantum Dynamics, 1040.

V. Rojansky, Introduction to Quantum Mechanics, 1950.

K. R. Dixit, The Elements of Wave Mechanics and Quantum Mechanics, 1953.

G. K. T. Conn, The Wave Nature of the Electron, 1944.

R. Beeching, Electron Diffraction, 1946.

G. T. Thomson and W. Cochrane, Theory and Practice of Electron Diffraction,1939.

J. Rice, introduction to Statistical Mechanics J or Students of Physics and Physical Chemistry, 1930.

E. H. Kennard, Kinetic Theory of Gases, 1938.

M. N. Saha and B. N. Srivastava, A Treatise on Heat, 1953.

N. Bohr, The Theory or Spectra and Atomic Constitution, '1924.

E. C. Stoner, Magnetism and Atomic Structure,• 1926.

A. Haas, Atomic. Theory, 1927.

A. Sommerfeld, Atomic Structure and Spectral Lines, 1934.

G. Herzberg, Atomic Spectra and Atomic Structure, 1937.

H. E. White, introduction to Atomic Spectra, 1934.

F. K. Richtmyer, Introduction to Modern Physics, 1934.

0. L. Padding and S. Goudsmit, The Structure of Line Spectra, 1930.

G. K. T. Conn, The Nature of the Atom, 1944.

F. 0. Rice and E. Teller, The Structure of Matter, 1949.

S. Tolansky, High Resolution Spectroscopy, 1947.

L. Kroenig, Bond Spectra and Molecular Spectra, 1930.

R. C. Johnson, Introduction to Molecular Spectra, 1949.

P. Debye, The Structure of Molecules, 1932.

K. B. Ramanathan, Molecular Scattering of Light, 1923.

8. Bhagavantham, Scattering of Light and Raman Effect, 1940.

G. Herzberg, Infra-red and Raman Spectra of Polyatomic Molecules, 1945.

G. B. B. Sutherland, Infra-red and Raman Spectra, 1935.

K. K. Darrow, Bell Telephone Series XII, Radioactivity, 1927.

F. Rasetti, Elements of Nuclear Physics, 19 7.

G. Gamow, Atomic Nuclei and Nuclear Transformations, 1937.

J. M. Cork, Radioactivity and Nuclear Physics, 1947.

K. K. Darrow, Bell Telephone Series XXII, Transmutation, 1931.

N. Feather, An Introduction to Nuclear Physics, 1936.

Reviews of Modern Physics, Nuclear Physics, C, July, 1937,

W. B. Mann, The Cyclotron, 1940.

D. H. Wilkinson. Ionizations Chambers and Counters, 1950.

J, B. Birks, Scintillation Counters, 1. 954.

S. C. Curran, Luminescence and the Scintillation Counter, 1953.

Curran and Craggs, Counting Tubes : Theory and Applications, 1949.

J. 0. Wilson, The Principles of Cloud Chamber Technique, 1951.

C. F. Powell and G. P. S. Occhialini, Nuclear Physics in Photographs, 1947.

H. Yagoda, Radioactive Measurements with Nuclear Emulsion, 1949.

J. B. Hoag and S. A. Kor, The Electron and Nuclear Physics, 1949-

P. B. Moon, Artificial Radioactivity, 1949.

E. Pollard and W. L. Davidson, Applied Nuclear Physics, 1946.

J. B. Rajam, Nuclear Isomerism (Thesis), 1939.

K. Mendelssohn, What is Atomic Energy ? 1946.

J. De Ment and H. C. Dake, Uranium and Atomic Power, 1945.

J. J. O'Neill, The Almighty' Atom, 1945.

G. Gamow, Atomic Energy, 1947.

W. E. Stephens, Nuclear Fission and Atomic Energy, 1948.

J. Cockcroft, Development and Future of Atomic Energy, 1950.

W. L. Lawrence, The Hell Bomb, 1951.

R. A. Millkan, Cosmic Rays,

L..Janossy, Cosmic Rays, 1948.

J. G. Wilson, About Cosmic Rays, 1948.

D. J. X. Montgomery, Cosmic Rays Physics, 1949.

L. Leprince Ringuet, Cosmic Rays, 1950.

B. Rossi, High-Energy Particles, 1952.

F. C. Frank and D. R. Rexworthy, Cosmic Radiation, 194.9.

A. Dauvillier, Les Rayons Cosmiques, 1954.

R. Marshals, Meson Physics, 1952.

H. A. Bathe and F. De Hoffman, Mesons and Fields, 1955.

J. G. Wilson, Progress in Cosmic Ray Physics, Vols. I (1952), II (1954) and III (1956).

Reviews of Modern Physics. Nuclear Physics, A and B, April, 1936 and April, 1931.

K. K. Darrow, Bell Telephone Series, The Nucleus-I, II, III, 1V, 1933-35.

D. Halliday, Introductory Nuclear Physics, 1950.

F. Bitter, Nuclear Physics, 1950.

S. Tolansky, Hyperfine Structure in Line Spectra and Nuclear Spin, 1948.

N. F. Ramsey, Nuclear Moments, 1953.

Reviews of Modern Physics, Jul?, 1946.

S. Devons, Excited States of Nuclei, 1949.

0. R. Frisch, Progress in Nuclear Physics, 1950.

F. Fermi, Nuclear Physics, 1951,

W. Heisenberg, Nuclear Physics, 1952.

J. M. Blatt and V. F. Weisskopi, Theoretical Nuclear Physics, 1952.

H. I. Bathe, _Elementary Nuclear Theory, 19 7.

C. Wentzel, Quantum Theory of Fields, 1949.

W. Pauli, Meson Theory of Nuclear Forces, 1948.

R. E. Marshak, Meson Physics, 1952.

R. D. Evans, The Atomic Nucleus, 1955.